Geometry

Grades 6-8

by

**Tiffany Moore and
Jenae Hawkins**

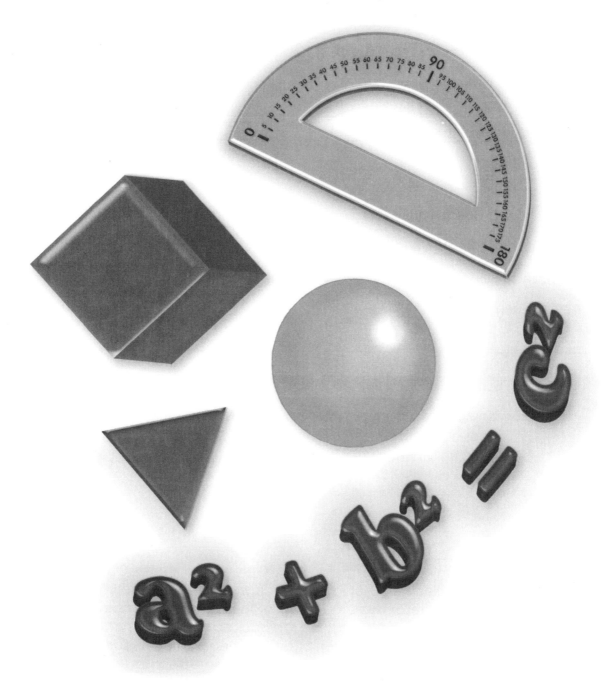

Carson-Dellosa Publishing Company, Inc.   •   Greensboro, North Carolina

Credits

Editors: Barrie Hoople and Joey Bland

Inside Illustrations: Van Harris and Lori Jackson

Cover Design: Lori Jackson

Content Review: Jen Bonnett

This book has been correlated to state, national, and Canadian provincial standards. Visit *www.carsondellosa.com* to search for and view its correlations to your standards.

ISBN 978-1-60022-530-7

Table of Contents

Introduction

The main objective of *Geometry 6–8* is to give students focused, grade-level appropriate practice to help them develop and reinforce geometry skills. To aid in this experience, the book offers an explanation of each individual skill followed by a variety of activities. These activities will ensure a greater understanding of each skill introduced. Review and critical thinking pages will challenge students to evaluate and apply each lesson and to consider real-life geometry applications.

Geometry 6–8 is divided into five sections. Each section is designed to lead students through the fundamentals of a skill to a challenging review. The concepts covered in this book include measuring angles; identifying polygons; calculating area, surface area, perimeter, and volume; transformations; coordinate graphing, and more. Included on pages 113–115 is a list and explanation of common geometric formulas and rules. A glossary of geometric terms is provided on pages 116–118

The activities in *Geometry 6–8* are a great way to challenge advanced students and to aid those in need of extra practice. Either focus for this book will yield the same result—an increased interest and understanding of valuable geometric concepts. Observe as your students experience how stimulating geometry can be!

Identifying Lines and Parts of Lines   Points, Lines, and Angles

A **point** is a position in a plane or in space that has no dimensions. These points are named, or written, points A, B, and C, or point A, point B, and point C.

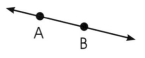

A **line** is a set of points in a straight path that extends infinitely in both directions. This line is named $\overleftrightarrow{AB}$ or $\overleftrightarrow{BA}$. Any two points on a line may be used to name it.

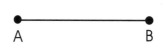

A **line segment** is a finite portion of a line that has two endpoints. This line segment is named $\overline{AB}$ or $\overline{BA}$. A segment must be named by its two endpoints.

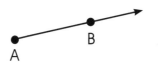

A **ray** is a portion of a line that extends from one endpoint infinitely in one direction. This ray is named $\overrightarrow{AB}$. The endpoint of a ray is written first, and any point on the ray may be used next.

Name each point, line, line segment, or ray.

1.

_____

2.

_____

3.

S   T

_____

4.

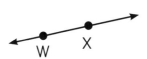

_____

5.

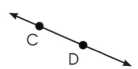

_____

6.

_____

7.

_____

8.

R   O

V

_____

9.

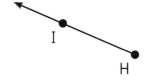

_____

Name: _____ Date: _____

Drawing Lines and Parts of Lines   Points, Lines, and Angles

Draw and label each of the following.

1. $\overleftrightarrow{AB}$

2. points C and D

3. $\overline{RS}$

4. points L, M, and N

5. $\overrightarrow{MN}$

6. $\overleftrightarrow{JK}$

Use the figure to the right to answer each question.

7. Name four points. _____

8. Name two line segments. _____

9. Name three rays. _____

10. Name the line three ways. _____

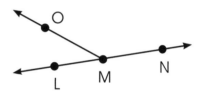

Use the figure to the right to answer each question.

11. Name five points. _____

12. Name two line segments. _____

13. Name four rays. _____

14. Name the line three ways. _____

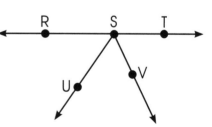

Intersecting and Parallel Lines
Points, Lines, and Angles

Intersecting lines are lines that cross each other at exactly one point, called the **point of intersection**.

Point X is the point of intersection of $\overleftrightarrow{AB}$ and $\overleftrightarrow{CD}$.

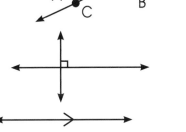

Perpendicular lines are two lines that form a right angle at their point of intersection.

Parallel lines are two lines that never intersect.

Identify each figure as parallel or perpendicular.

1.

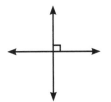

_____

2.

_____

3.

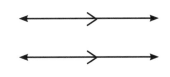

_____

Use the figure to the right to answer each question.

4. Name the point of intersection. _____

5. Name the two lines that intersect. _____

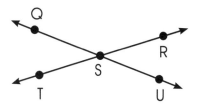

Draw and label each of the following.

6. $\overleftrightarrow{LM}$ intersects $\overrightarrow{NO}$ at point P

7. Y is the point at which $\overleftrightarrow{XZ}$ intersects $\overleftrightarrow{WV}$

8. $\overleftrightarrow{HI}$ is perpendicular to $\overleftrightarrow{JK}$

9. $\overleftrightarrow{RS}$ is parallel to $\overleftrightarrow{TU}$

Name: _____   Date: _____

Midpoints
Points, Lines, and Angles

A **midpoint** is a point that bisects, or divides, a line segment into two **congruent** parts.

Point B is the midpoint of $\overline{AC}$. Therefore, $\overline{AB} \cong \overline{BC}$.

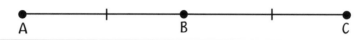

Name the midpoints.

1.    2.   3.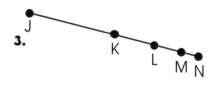

_____   _____   _____

Draw and label each of the following.

4. L is the midpoint of $\overline{KM}$.   **5.** $\overline{CD}$ is bisected by point M.   **6.** $\overline{XY} \cong \overline{YZ}$, on $\overline{XZ}$.

Use the figure and information below to find the length of each segment.

D is the midpoint of $\overline{CE}$.

B is the midpoint of $\overline{AC}$.

C is the midpoint of $\overline{AE}$.

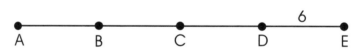

7. $\overline{AB} =$ _____   **8.** $\overline{BC} =$ _____   **9.** $\overline{AC} =$ _____

10. $\overline{CD} =$ _____   **11.** $\overline{CE} =$ _____   **12.** $\overline{AE} =$ _____

Use the figure and information below to find the length of each segment.

K bisects $\overline{JM}$.

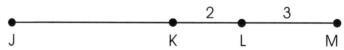

13. $\overline{KM} =$ _____   **14.** $\overline{JK} =$ _____   **15.** $\overline{JM} =$ _____

Name: _____ Date: _____

Opposite Rays
Points, Lines, and Angles

Linear—something that relates to, or resembles, a line

Opposite Rays are two rays that share an endpoint and extend in opposite directions to form a line. $\overrightarrow{BA}$ and $\overrightarrow{BC}$ are opposite rays.

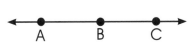

Use the figure to the right to answer each question.

1. Name the opposite rays. _____

2. Name the line formed by the opposite rays. _____

Use the figure to the right to answer each question.

3. Name the opposite rays. _____

4. Name the line formed by the opposite rays. _____

Name the pair of opposite rays in each figure.

5.
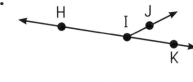

_____

6.
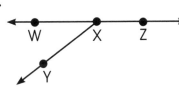

_____

7.
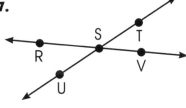

_____

_____

8.
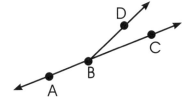

_____

9.
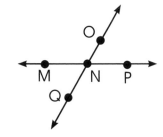

_____

_____

10.
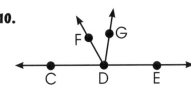

_____

Collinear Points

Points, Lines, and Angles

When three or more points lie on the same line, they are **collinear**.

Points A, B, and C are collinear.

Points X, Y, and Z are not collinear.

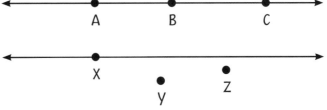

For each statement, circle *True* or *False*. Then, explain your answer

1. Points A, B, and C are collinear.   True   False _____

_____

2. Points W, X, Y, and Z are collinear.   True   False _____

_____

3. Points J, K, and L are not collinear.   True   False _____

_____

4. Points S and T are collinear.   True   False _____

Use the figure to the right to answer each question.

5. Name all sets of collinear points.

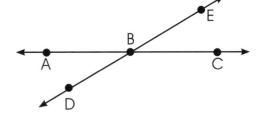

_____

6. Name the points that are not collinear to $\overleftrightarrow{DE}$. _____

Use the provided information to draw and label the figure in the box to the right.

7. $\overleftrightarrow{BK}$ intersects $\overleftrightarrow{JM}$ at point L.
$\overleftrightarrow{RT}$ is parallel to $\overleftrightarrow{JM}$ and intersects $\overleftrightarrow{BK}$ at point C.

Use your drawing to name all sets of collinear points and two sets of points that are not collinear.

Collinear _____

Not collinear _____

Name: _____ Date: _____

Identifying Planes

Points, Lines, and Angles

A **plane** is a flat surface that extends infinitely in all directions. Three points that are not collinear are needed to determine a plane.

When three or more points that are not collinear lie in the same plane, they are **coplanar**.

Points are coplanar.

Points are not coplanar.

Decide whether each set of points determines a plane. Circle *Yes* or *No*.

1.

• A   • B

Yes   No

2.

• C   • E
   • D

Yes   No

3.

• F   • G   • H

Yes   No

4.

• I

Yes   No

Identify the points in each figure as coplanar or not coplanar.

5.

_____

6.

_____

7.

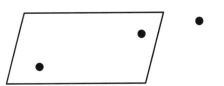

_____

8.

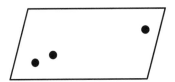

_____

Coplanar Points

Points, Lines, and Angles

Flat-Plane Rule

If three points are coplanar, then the line containing two of the points is in the same plane.

Points A and B cannot form a line in plane L because they are not coplanar.

Points X and Y can form a line in plane M because they are coplanar.

Plane L

●D ●C   ●B

●A

Plane M

●
Y

●   ●
X   Z

Determine whether the points can form a line in the given plane. Then, write *Yes* or *No*.

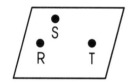

1.

Points R and S _____

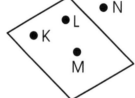

2.

Points M and N _____

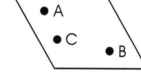

3.

Points A and B _____

Draw and label each of the following.

4. Three collinear points in plane A

5. Three coplanar points in plane C

6. $\overleftrightarrow{LN}$ exists in plane M

7. One point that is not coplanar to plane S

8. Points D, E, F are collinear in plane A, but not in plane B

9. Collinear points Q, R, and S in plane F

Critical Thinking   Points, Lines, and Angles

Answer each question. Show your work. Hint: Draw pictures to help.

1. $\overleftrightarrow{AB}$ bisects $\overleftrightarrow{CD}$ at point E. $\overline{CE}$ measures 12 m. How long is $\overline{CD}$? _____

2. $\overrightarrow{IH}$ and $\overrightarrow{IJ}$ are opposite rays. Name the line formed three ways. _____

Use the following information to draw and label a figure below. Then, answer the questions.

3. $\overleftrightarrow{RS}$ intersects $\overleftrightarrow{TU}$ at point V. $\overleftrightarrow{RS}$ is perpendicular to $\overleftrightarrow{TU}$. $\overleftrightarrow{WX}$ is parallel to $\overleftrightarrow{RS}$ and intersects $\overleftrightarrow{TU}$ at point Y.

Is $\overleftrightarrow{WX}$ perpendicular to $\overleftrightarrow{TU}$? _____

Name all sets of collinear points. _____

4. Draw and label the following. Points M, N, and O lie in plane M.

Are points M, N, and O coplanar? ____

Review   Points, Lines, and Angles

Write the letter for the correct term beside each definition.

1. _____ The point at which two lines intersect
2. _____ A set of points in a straight path that extends infinitely in both directions
3. _____ Two lines that form a right angle at their point of intersection
4. _____ Position in space, often represented by a dot
5. _____ A finite portion of a line that has two endpoints
6. _____ Three or more points that lie in the same line
7. _____ A point that bisects a line segment
8. _____ Lines in the same plane that never intersect
9. _____ A portion of a line that extends from one endpoint infinitely in one direction
10. _____ A flat surface that extends infinitely in all directions
11. _____ Two rays that share an endpoint and extend in opposite directions to form a line
12. _____ Something that relates to or resembles a line
13. _____ Three or more points that lie in the same plane
14. _____ If three points are coplanar, then the line containing two of the points is in the same plane.

> A. opposite rays
> B. point
> C. ray
> D. point of intersection
> E. linear
> F. midpoint
> G. parallel lines
> H. collinear points
> I. perpendicular lines
> J. line segment
> K. line
> L. plane
> M. coplanar points
> N. flat-plane rule

Write the letter for the correct term beside each diagram.

15. _____

16. _____

17. _____

18. _____

19. _____

20. _____

> A. intersecting lines
> B. line
> C. line segment
> D. parallel lines
> E. coplanar points
> F. collinear points

Name: _____ Date: _____

<table>
<tr><td>Review</td><td>Points, Lines, and Angles</td></tr>
</table>

Use the diagram to the right to answer each question.

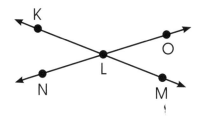

1. Name two lines. _____

2. Name four line segments. _____

3. Name four rays. _____

4. Name the point of intersection. _____

Use the diagram and information below to find the length of each segment.

T is the midpoint of $\overline{RV}$.

S is the midpoint of $\overline{RT}$.

U is the midpoint of $\overline{TV}$.

R   S   9   T   U   V

5. $\overline{RS}$ = _____

6. $\overline{RT}$ = _____

7. $\overline{TU}$ = _____

8. $\overline{UV}$ = _____

9. $\overline{TV}$ = _____

10. $\overline{RV}$ = _____

Use the diagram to the right to answer each question.

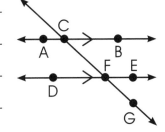

11. Name the parallel lines. _____

12. Name the two points of intersection. _____

13. Name the line that intersects $\overleftrightarrow{AB}$ and $\overleftrightarrow{DE}$. _____

14. Name all sets of collinear points. _____

Answer the question about each diagram.

15.

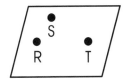

Are these points coplanar? _____

16.

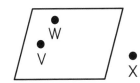

Are these points coplanar? _____

Name: _____ Date: _____

Identifying Angles
Points, Lines, and Angles

Angle—two rays that share an endpoint

Vertex—point where two segments, lines, or rays meet to form an angle

Acute Angle—an angle with a measure greater than 0° but less than 90°

Obtuse Angle—an angle with a measure greater than 90° but less than 180°

Right Angle—an angle with a measure equal to 90°

Straight Angle—an angle with a measure equal to 180°, or a straight line

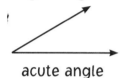
acute angle

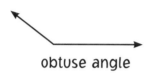
obtuse angle

right angle

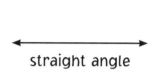
straight angle

An angle is named by its vertex or with an endpoint first, the vertex in the middle, and the other endpoint last.

Identify each angle as acute, obtuse, right, or straight.

1. m∠RST = 180° _____

2. m∠ABC = 45° _____

3. m∠TUV = 90° _____

4. m∠XYZ = 135° _____

5. m∠IJK = 100° _____

6. m∠GHI = 80° _____

Identify each diagram as acute, obtuse, right, or straight.

7.

_____

8.
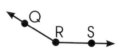
_____

9.

_____

10.

_____

11.
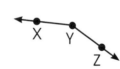
_____

12.
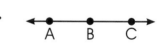
_____

Draw and label each of the following.

13. Acute angle HIJ

14. Obtuse angle DEF

15. Right angle GHI

Naming and Classifying Angles   Points, Lines, and Angles

An angle is named by its vertex, or with an endpoint first, the vertex in the middle, and the other endpoint last.

This angle is named ∠Y, ∠XYZ, or ∠ZYX.

$\overrightarrow{YX}$ and $\overrightarrow{YZ}$ are sides, and point Y is the vertex.

Adjacent Angles—two angles that have a side and vertex in common ∠DBC and ∠ABC are adjacent angles.

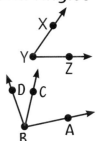

Name each angle.

1.

_____

2.

_____

3.

_____

Name the vertex and sides of each angle below.

4.

vertex _____

sides _____

5.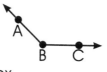

vertex _____

sides _____

6.

vertex _____

sides _____

7.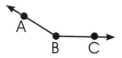

vertex _____

sides _____

8.

vertex _____

sides _____

9.

vertex _____

sides _____

Use the diagrams below to answer each question.

10. Name the obtuse angle's sides. _____

11. Name the vertex. _____

12. Are ∠NMO and ∠LMO adjacent? _____

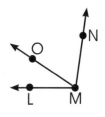

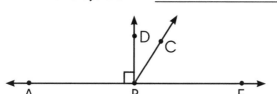

13. Are ∠ABC and ∠CBE adjacent? _____ If not, explain. _____

_____

14. Are ∠DBE and ∠ABC adjacent? _____ If not, explain. _____

_____

Linear Pairs
Points, Lines, and Angles

Linear Pair—a pair of adjacent angles that forms when two lines intersect. ∠ABD and ∠CBD, ∠CBD and ∠CBE, ∠CBE and ∠EBA, and ∠EBA and ∠ABD form linear pairs.

Two angles are **supplementary** if the sum of their measures is 180°.

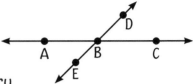

Linear Pair Rule
If two angles form a linear pair, then they are supplementary.
∠ABD and ∠CBD, ∠CBD and ∠CBE, ∠CBE and ∠EBA, and ∠EBA and ∠ABD form linear pairs, so they are supplementary.

Fill in each blank with the correct answer.

1. ∠HIJ and ∠JIK form a _____.

2. ∠DEG and ∠ ____ form a linear pair.

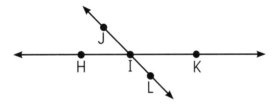

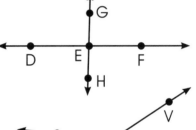

3. Name each linear pair in the diagram to the right.

∠ _____ and ∠ _____; ∠ _____ and ∠ _____;

∠ _____ and ∠ _____; ∠ _____ and ∠ _____

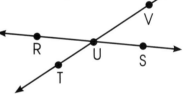

Find each missing angle measure.

4.

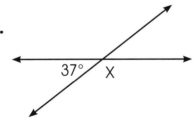

m∠X = _____

5.

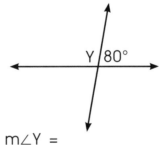

m∠Y = _____

6.

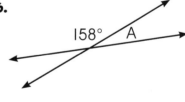

m∠A = _____

7.

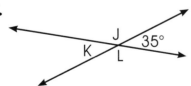

m∠J = _____
m∠K = _____
m∠L = _____

8.

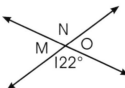

m∠M = _____
m∠N = _____
m∠O = _____

9.

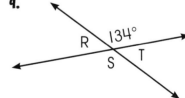

m∠R = _____
m∠S = _____
m∠T = _____

Name: _____ Date: _____

Use the diagram to the right to answer each question.

1. m∠A + m∠B = _____

These are called _____ angles.

2. m∠D + m∠_____ = 180°

These are called _____ angles.

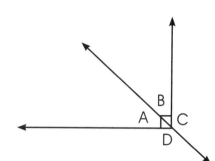

Use the diagram to the right to answer each question.

3. If m∠H = 43°,

m∠E = _____

m∠G = _____

m∠F = _____

4. If m∠G = 132°,

m∠H = _____

m∠E = _____

m∠F = _____

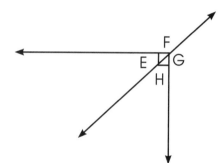

Use the diagram to the right to answer each question.

5. m∠R + m∠ T = _____

6. m∠Q + m∠S = _____

7. m∠S + m∠ T = _____

8. If m∠S = 32°, m∠T = _____ m∠R = _____ m∠Q = _____

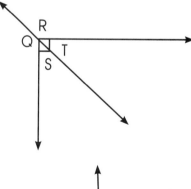

Use the diagram to the right to answer each question.

9. If m∠X = 45°, m∠R = _____ m∠Y = _____ m∠Z = _____

10. m∠X + m∠Y = _____

11. m∠R + m∠X = _____

12. ∠R + ∠X are _____ .

Interior, Exterior, and Corresponding Angles Points, Lines, and Angles

Transversal—a line that intersects two parallel lines to form eight angles

Alternate Exterior Angles—Alternate exterior angles are pairs of angles that lie outside the parallel lines on opposite sides of the transversal. Angles A and H and angles B and G are alternate exterior angles.

Alternate Interior Angles—Alternate interior angles are pairs of angles that lie between the parallel lines on opposite sides of the transversal. Angles C and F and angles D and E are alternate interior angles.

Consecutive Interior Angles—Consecutive interior angles are pairs of angles that lie between the parallel lines on the same side of the transversal. Angles C and E and angles D and F are consecutive interior angles.

Corresponding Angles—Corresponding angles are pairs of angles that appear in corresponding positions in the two sets of angles formed by parallel lines. Angles A and E, angles B and F, angles C and G, and angles D and H are corresponding angles.

Use the diagram to the right to list all of the pairs of angles that fit each description.

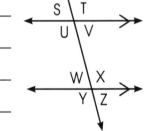

1. alternate exterior angles _____

2. alternate interior angles _____

3. consecutive interior angles _____

4. corresponding angles _____

_____

Use the diagram to the right to identify each pair of angles as alternate exterior, alternate interior, corresponding, or consecutive interior.

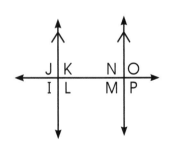

5. ∠I and ∠M are _____ angles.

6. ∠J and ∠P are _____ angles.

7. ∠K and ∠M are _____ angles.

8. ∠L and ∠M are _____ angles.

9. ∠I and ∠O are _____ angles.

10. ∠K and ∠O are _____ angles.

Parallel Lines Cut by a Transversal   Points, Lines, and Angles

When parallel lines are cut by a transversal, the following rules are true:

Rule 1: Alternate exterior angles are congruent. Their angles have the
same measure

$\angle A \cong \angle H$ and $\angle B \cong \angle G$

Rule 2: Alternate interior angles are congruent.

$\angle C \cong \angle F$ and $\angle D \cong \angle E$

Rule 3: Corresponding angles are congruent.

$\angle A \cong \angle E$ $\angle B \cong \angle F$ $\angle C \cong \angle G$ $\angle D \cong \angle H$

Rule 4: Consecutive interior angles are supplementary.

Angles C and E and angles D and F are supplementary.

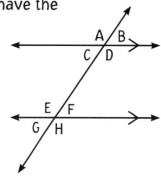

Use the diagram to the right to answer each question.

1. $\angle I \cong \angle$ _____ and $\angle$ _____

2. $\angle J \cong \angle$ _____ and $\angle$ _____

3. $\angle J$ and $\angle$ _____ are supplementary (using rule 4).

4. $\angle L \cong \angle$ _____ and $\angle$ _____

5. $\angle O$ and $\angle$ _____ are supplementary (using rule 4).

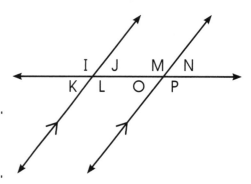

Use the diagram to the right to answer each question, then state the rules used.

6. a. $\angle S \cong$ _____

   _____

b. $\angle W \cong$ _____

   _____

c. $\angle T \cong$ _____

   _____

d. $\angle X \cong$ _____

   _____

e. $\angle U \cong$ _____

   _____

f. $\angle Y \cong$ _____

   _____

g. $\angle V \cong$ _____

   _____

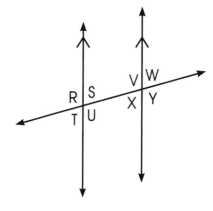

7. Are $\angle T$ and $\angle X$ supplementary? _____ If not, explain. _____

_____

8. If $m\angle R = 105°$, then $m\angle V =$ _____.

Using a Protractor   Points, Lines, and Angles

Use a protractor to draw and measure angles accurately.

How to Use a Protractor

1. Find the center along the straight edge of the bottom of the protractor. Place the center over the vertex of the angle you wish to measure.

2. Position the protractor so that the 0° mark on the straight edge lines up with one side of the angle.

3. Determine the type of angle and which set of numbers to use. Find the point where the second side of the angle intersects the numbered edge of the protractor. If the angle does not extend far enough to intersect the lines, use the protractor like a ruler to extend the side.

4. Read the number that is written where the side crosses the protractor. This is the measure of the angle in degrees.

Use a protractor to measure each specified angle to the nearest degree. Write the type (*right*, *obtuse*, or *acute*) and measure of each angle.

1.

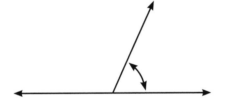

type:_____   measure: _____

2.

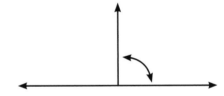

type:_____   measure: _____

3.

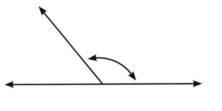

type:_____   measure: _____

4.

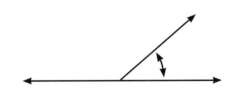

type:_____   measure: _____

5.

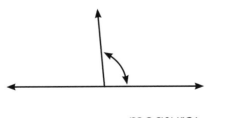

type:_____   measure: _____

6.

type:_____   measure: _____

Name: _____   Date: _____

Use the diagram to the right to answer each question.

1. ∠A ≅ ∠ _____ and ∠ _____

∠D ≅ ∠ _____ and ∠ _____

∠C and ∠ _____ are supplementary.

∠C ≅ ∠ _____ and ∠_____

∠F and ∠_____ are supplementary.

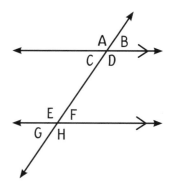

Use the diagram to the right to answer each question.

2. If m∠H = 143°,

m∠E = _____

m∠F = _____

m∠G = _____

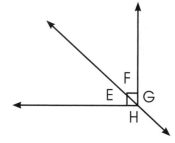

3. ∠LMO and ∠OMP form a linear pair.

If m∠LMO = 62°, m∠OMP = _____ .

4. Use a protractor to measure each specified angle to the nearest degree.

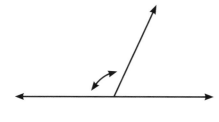

_____

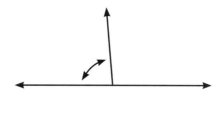

_____

5. Using a protractor, draw and label ∠TUV and ∠WUV. ∠TUV and ∠WUV form a linear pair. The measure of ∠TUV = 45°.

Name: _____ Date: _____

Write the letter for the correct term beside each definition.

1. _____ two angles in which measures add to 180°

2. _____ pairs of angles that lie outside parallel lines on opposite sides of the transversal

3. _____ an angle with a measure equal to 180°

4. _____ two rays that share an endpoint

5. _____ a point at which two line segments, lines, or rays meet

6. _____ an angle with a measure greater than 0° but less than 90°

7. _____ angles with the same measure

8. _____ two angles that have a vertex and side in common

9. _____ an angle with a measure greater than 90° but less than 180°

10. _____ pairs of angles that lie between parallel lines on opposite sides of the transversal

11. _____ two angles in which measures add to 90°

12. _____ an angle with a measure equal to 90°

A. angle

B. vertex

C. congruent angles

D. alternate interior angles

E. alternate exterior angles

F. acute angle

G. right angle

H. obtuse angle

I. straight angle

J. adjacent angles

K. complementary angles

L. supplementary angles

Identify each angle as right, obtuse, acute, or straight.

13. m∠T = 23° _____   **14.** m∠X = 91° _____   **15.** m∠N = 180° _____

Find each missing angle measure.

16.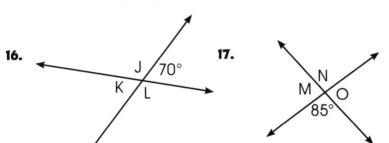

17.

18.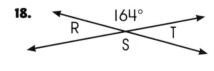

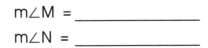

m∠J = _____   m∠M = _____   m∠R = _____

m∠K = _____   m∠N = _____   m∠S = _____

m∠L = _____   m∠O = _____   m∠T = _____

Name: _____ Date: _____

Review
Points, Lines, and Angles

Use the diagram to the right to identify each pair of angles as alternate exterior, alternate interior, or consecutive interior.

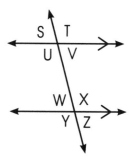

1. ∠U and ∠X are _____ angles.

2. ∠T and ∠Y are _____ angles.

3. ∠U and ∠W are _____ angles.

4. ∠S and ∠Z are _____ angles.

5. ∠V and ∠W are _____ angles.

6. ∠V and ∠X are _____ angles.

Use a protractor to measure each specified angle to the nearest degree. Write the type (*right*, *obtuse*, or *acute*) and measure of each angle.

7.

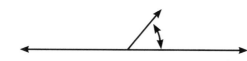

type:_____ measure:_____

8.

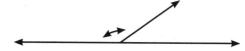

type:_____ measure:_____

9.

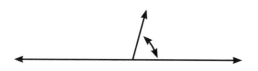

type:_____ measure:_____

10.

type:_____ measure:_____

11.

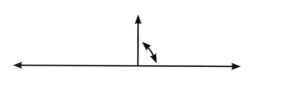

type:_____ measure:_____

12.

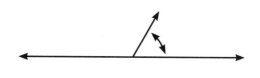

type:_____ measure:_____

Name: _____ Date: _____

Polygons
A **polygon** is a simple, closed plane figure formed by line segments that has two sides meeting at each vertex.

1. Which of the following shapes are polygons? _____

A.　　　　　B.　　　　　C.　　　　　D.　　　　　E.　　　　　F.

2. Three of the shapes above are not polygons. Explain why each one is not a polygon.

shape _____　_____

shape _____　_____

shape _____　_____

Complete the table.

Type of Polygon	Number of Sides
3.	3
4. quadrilateral	
5.	5
6. hexagon	
7. heptagon	
8.	8
9. nonagon	
10. decagon	

Use the table above to identify each polygon.

11. 　　　12. 　　　13.

_____　　_____　　_____

14. 　　　15. 　　　16.

_____　　_____　　_____

Name: _____   Date: _____

Identifying Polygons   Geometric Figures

Identify each figure and then write the number of sides.

	Name	**Number of Sides**

1.   _____   _____

2.   _____   _____

3.   _____   _____

4.   _____   _____

5.   _____   _____

6.   _____   _____

7.   _____   _____

8.   _____   _____

Name: _____   Date: _____

Identifying Regular Polygons   Geometric Figures

A polygon can be classified as regular or irregular. **Regular polygons** are equilateral, which means they have sides of equal length. They are also **equiangular**, which means they have angles of equal measure.

Equilateral 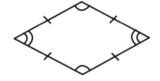   Equiangular

Hint: You can tell which sides and angles are congruent by the matching markings on the sides and angle measures. Remember, a polygon is regular only if it is both equilateral and equiangular.

Identify each polygon as equilateral, equiangular, regular, or none.

1.

_____

2.

_____

3.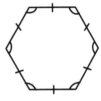

_____

4.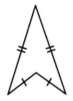

_____

5.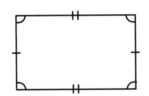

_____

6.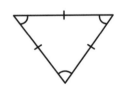

_____

7.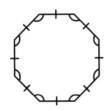

_____

8.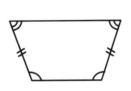

_____

9.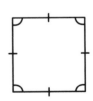

_____

10.

_____

11.

_____

12.

_____

Quadrilaterals and Parallelograms Geometric Figures

Quadrilateral—a four-sided polygon
Parallelogram—a quadrilateral in which opposite sides are parallel

Squares, rhombuses, and rectangles are parallelograms.

Use these rules to determine if a quadrilateral is a parallelogram. If any rule is proven, the figure is a parallelogram.
Rule 1: Both pairs of opposite sides are parallel.
Rule 2: Both pairs of opposite sides are congruent.
Rule 3: Both pairs of opposite angles are congruent.
Rule 4: The diagonals (segments that connect opposite vertices) bisect, or divide, each other into two congruent segments.
Rule 5: All consecutive angles (angles that share a side) are supplementary.

Use the information provided for each quadrilateral to write the number of the rule or rules that prove it is a parallelogram.

1.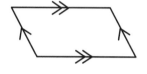

_____

_____

2.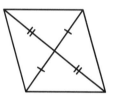

_____

_____

3.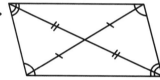

_____

_____

4.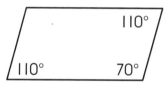

_____

_____

5.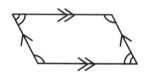

_____

_____

6.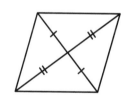

_____

_____

7.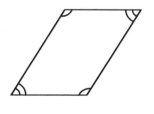

_____

_____

8.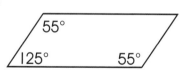

_____

_____

9.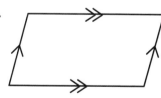

_____

_____

10.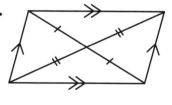

_____

_____

11.

_____

_____

12.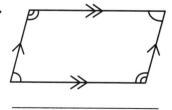

_____

_____

Name: _____   Date: _____

Identifying Parallelograms
Geometric Figures

To identify parallelograms, review the following terms.

Consecutive Angles—angles that share a side (∠ABC and ∠DAB, ∠DAB and ∠ADC, ∠ADC and ∠DCB, ∠DCB and ∠CBA); consecutive angles are supplementary

Diagonal—a line segment that connects two nonconsecutive, or opposite, vertices of a polygon ($\overline{BD}$ and $\overline{AC}$)

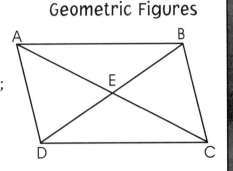

Determine whether each quadrilateral is a parallelogram. Write *Yes* or *No*. If the answer is no, explain why.

1.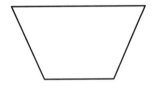

2.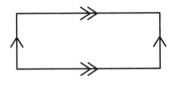

3.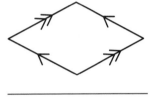

_____   _____   _____

_____   _____   _____

ABCD is a parallelogram. Find each missing side length and explain how you found the answer.

4. $\overline{AB}$ = _____   Explanation: _____

_____

5. $\overline{DA}$ = _____   Explanation: _____

_____

6. $\overline{DE}$ = _____   Explanation: _____

_____

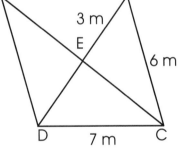

EFGH is a parallelogram. Find each missing angle measure and explain how you found the answer.

7. m∠FGH = _____   Explanation: _____

_____

8. m∠GHE = _____   Explanation: _____

_____

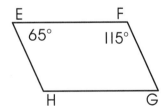

LMNO is a parallelogram. Answer each question and explain how you found the answer.

9. m∠LMN + m∠MNO = _____ Explanation: _____

_____

10. Is $\overline{LP}$ ≅ $\overline{NP}$? _____ Explanation: _____

_____

Parallelogram Practice Geometric Figures

Use parallelogram PQRS to answer each question.

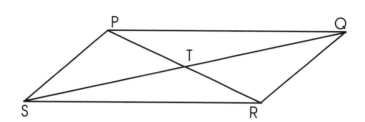

1. $\overline{SP} \cong$ _____

2. $\angle RQP \cong$ _____

3. $\overline{ST} \cong$ _____

4. $\overline{PT} \cong$ _____

5. $\angle SPQ \cong$ _____

6. $\overline{PQ} \cong$ _____

Find each measurement in parallelogram ABCD.

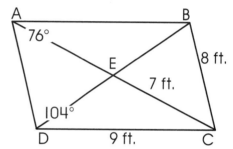

7. $\overline{AB} =$ _____   **8.** $\overline{AD} =$ _____

9. $\overline{AE} =$ _____   **10.** $\overline{AC} =$ _____

11. $m\angle ABC =$ _____   **12.** $m\angle BCD =$ _____

Use parallelogram DEFG to answer each question.

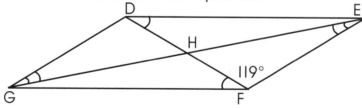

13. If $\overline{DH} = 7.5$, $\overline{FH} =$ _____ $\overline{DF} =$ _____

14. If $m\angle FDE = 30°$, $m\angle DFG =$ _____ $m\angle EFG =$ _____

Circle the correct answer for each question.

15. The diagonals of a parallelogram must_____.

    a. bisect each other   b. be parallel   c. be perpendicular   d. be congruent

16. Which statement is NOT true about parallelograms?

    a. Opposite sides are parallel.   b. Opposite sides are congruent.

    c. Opposite angles are supplementary.   d. Opposite angles are congruent.

Convex and Concave Polygons   Geometric Figures

One way to classify polygons is to determine whether they are convex or concave.
Convex polygons contain angles in which measures are less than 180°. No lines
intersect within the figure. **Concave polygons** contain at least one interior angle with a
measure greater than 180°. Also, lines drawn along a concave polygon's sides intersect
within the figure.

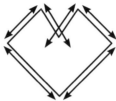

Convex   Concave

Identify each polygon as convex or concave.

1.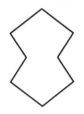

_____

2.

_____

3.

_____

4.

_____

5.

_____

6.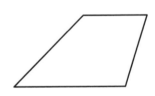

_____

7.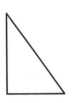

_____

8.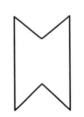

_____

9.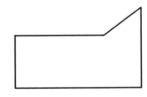

_____

10.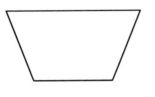

_____

11.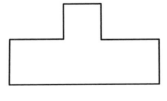

_____

12.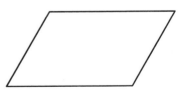

_____

Identifying Solids

Geometric Figures

Solid—a closed, three-dimensional figure that contains edges, faces, and vertices

Edge—a line segment where two faces in a solid meet

Face—a shape bounded by edges in a solid

Vertex—a point on a solid where three or more faces intersect

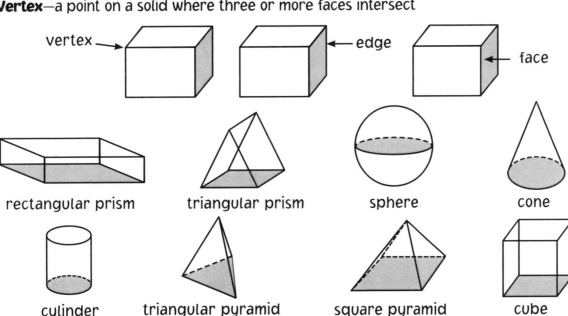

vertex → ← edge ← face

rectangular prism   triangular prism   sphere   cone

cylinder   triangular pyramid   square pyramid   cube

Identify each solid by writing *triangular pyramid, cube, triangular prism, square pyramid,* or *sphere*.

1. 6 faces, 12 edges, 8 vertices _____

2. 4 faces, 6 edges, 4 vertices _____

3. 5 faces, 9 edges, 6 vertices _____

4. 0 faces, 0 edges, 0 vertices _____

5. 5 faces, 8 edges, 5 vertices _____

Draw each solid.

6. rectangular prism   **7.** cylinder   **8.** cube

Name: _____ Date: _____

Exploring Solids

Use a word from the box below to complete each statement or question.

one	six	rectangles	squares	sphere

1. All of the faces of a cube are _____ .

2. All of the faces of a rectangular prism are _____ .

3. How many faces are on a rectangular prism? _____

4. How many vertices does a cone have? _____

5. What solid has neither vertices nor edges? _____

Complete the table.

	Name	Number of Faces	Number of Edges	Number of Vertices
6.	rectangular Prism			
7.	sphere			
8.	cone			
9.	cylinder			
10.	square pyramid			
11.	cube			

Write the name of the correct solid for each definition.

12. _____ A solid with one vertex joined to a circular base

13. _____ A solid that has congruent bases, parallel circles

14. _____ A solid with a triangular base and triangular faces that meet at a common vertex

15. _____ A solid in which every face is a square

16. _____ A solid in which all points on the surface are equidistant from the center

Name: _____ Date: _____

Critical Thinking   Geometric Figures

Use a term from the box below to answer each question.

| triangular prism | cube | square pyramid | cylinder |
| triangular pyramid | cone | rectangular prism | sphere |

1. I have 2 triangular faces.

I have 3 rectangular faces

What am I? _____

2. I have 8 vertices.

I have rectangular faces.

What am I? _____

3. I have no flat faces.

I can roll.

What am I? _____

4. I have 4 faces.

I have 4 vertices.

What am I? _____

5. I have 6 square faces.

I have 12 edges.

What am I? _____

6. I have 2 circular faces.

I have a curved surface.

What am I? _____

7. I have 1 vertex.

I have 1 circular face.

What am I? _____

8. I have 1 square face.

I have 4 triangular faces.

What am I? _____

Name: _____ Date: _____

Review   Geometric Figures

Identify each polygon as regular or irregular.

1.

_____

2.

_____

3.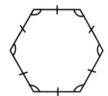

_____

Use parallelogram ABCD to answer each question.

4. $\overline{DA} \cong$ _____

5. $\angle CBA \cong$ _____

6. $\overline{DE} \cong$ _____

7. $\overline{AE} \cong$ _____

8. $\angle DAB \cong$ _____

9. $\overline{AB} \cong$ _____

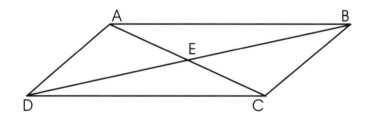

Identify each polygon below as convex or concave.

10.

_____

11.

_____

12.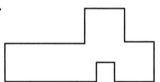

_____

13.

_____

14.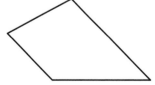

_____

15.

_____

Name: _____ Date: _____

Review

Use parallelogram ABCD to answer each question.

1. If m∠BAD = 32°, find the measures of the other
three angles of ABCD.

m∠BCD = _____

m∠ADC = _____

m∠ABC = _____

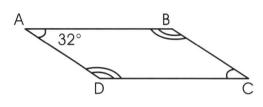

Use parallelogram VWXY to answer each question.

2. $\overline{VW} \cong$ _____

3. $\overline{XW} \cong$ _____

4. $\overline{VZ} \cong$ _____

5. ∠VWX ≅ _____

6. $\overline{YZ} \cong$ _____

7. ∠WXY ≅ _____

8. Point Z is the midpoint of _____ and _____ .

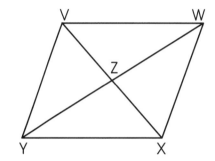

Identify each solid.

9.

_____

10.

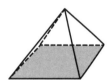

_____

11.

_____

Draw each solid.

12. prism

13. cylinder

14. cone

Identifying Triangles
Geometric Figures

A **triangle** is a three-sided polygon. A triangle's angles and sides can be used to classify it.

Classifying by Angles

Acute Triangle	Equiangular Triangle	Right Triangle	Obtuse Triangle
		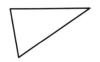	
3 acute angles	3 congruent angles	1 right angle	1 obtuse angle

Classifying by Sides

Equilateral Triangle	Isosceles Triangle	Scalene Triangle

3 congruent sides	2 congruent sides	0 congruent sides

Classify each triangle by its angles and sides.

1.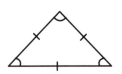

_____
_____

2.

_____
_____

3.

_____
_____

4.

_____
_____

5.

_____
_____

6.

_____
_____

7.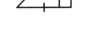

_____
_____

8.

_____
_____

9.

_____
_____

Name: _____ Date: _____

Classifying Triangles   Geometric Figures

The sum of the measures of a triangle's angles is 180°.

Write the letter for the correct term beside each triangle description.

1. Side lengths are 3 cm, 4 cm, and 6 cm. _____

2. Angle measures are 40°, 60°, and 80°. _____

3. Angle measures are 25°, 10°, and 145°. _____

4. Side measures are 6 in., 5 in., and 6 in. _____

5. Angle measures are 30°, 60°, and 90°. _____

A. right

B. obtuse

C. acute

D. scalene

E. isosceles

Complete each statement using *sometimes*, *always*, or *never*.

6. A triangle _____ has a right angle and an obtuse angle.

7. The acute angles of a right triangle are _____ complementary.

8. An obtuse triangle is _____ an isosceles triangle.

9. An isosceles triangle is _____ an equilateral triangle.

10. A triangle's angles _____ add to 180°.

Circle the correct term below each statement.

11. A triangle with two 45° angles

  a. right   b. acute   c. equiangular   d. obtuse   e. scalene

12. A triangle with a 102° angle

  a. right   b. acute   c. equiangular   d. obtuse   e. scalene

13. A triangle in which all angles measure 60°

  a. right   b. acute   c. equiangular   d. obtuse   e. scalene

14. A triangle in which side measures are 2 m, 4 m, and 7 m

  a. right   b. acute   c. equiangular   d. obtuse   e. scalene

15. A triangle in which angle measures are 45°, 55°, and 80°

  a. right   b. acute   c. equiangular   d. obtuse   e. scalene

Parts of Triangles   Geometric Figures

Triangles are made of **sides** (the edges or boundaries of the triangle) and **vertices** (where two sides join). Two sides that have a common vertex are called **adjacent sides**.

In a right triangle, the sides adjacent to the right angle are called **legs** and the side opposite the right angle is called the **hypotenuse**.

In an isosceles triangle, the congruent sides are called **legs** and the bottom side is called the **base**.

Identify the specified part(s) of each triangle below. Write *legs*, *vertex*, *adjacent sides*, *hypotenuse*, or *base*.

1.

_____

2.
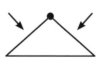
_____

3.

_____

4.

_____

5.
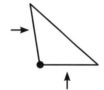
_____

6.

_____

7.
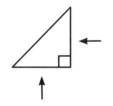
_____

8.
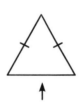
_____

9.

_____

10.

_____

11.

_____

12.
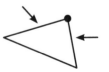
_____

Name: _____ Date: _____

Identifying Congruent Triangles

Geometric Figures

Angle-Side-Angle Rule (ASA)
Two triangles are congruent if two angles and the side between those angles of one triangle are congruent to the corresponding parts of the other triangle.

Angle-Angle-Side Rule (AAS)
Two triangles are congruent if two angles and a side opposite one of those angles in one triangle are congruent to the corresponding parts of the other triangle.

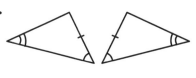

Side-Side-Side Rule (SSS)
Two triangles are congruent if three sides of one triangle are congruent to the corresponding sides of the other triangle.

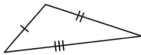 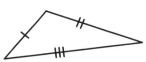

Side-Angle-Side Rule (SAS)
Two triangles are congruent if two sides and the angle between those sides of one triangle are congruent to the corresponding parts of the other triangle.

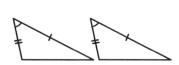

Determine which rule is used to prove that the triangles below are congruent. Write *ASA*, *SSS*, *AAS*, or *SAS* under each pair of triangles.

1.

_____

2.

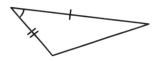

_____

3.

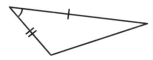

_____

4.

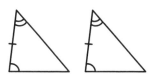

_____

5.

_____

6.

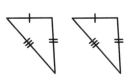

_____

7.

_____

8.

_____

9.

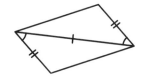

_____

Congruent Triangles

Geometric Figures

Rule 1
If two sides of a triangle are congruent, then the angles opposite them are also congruent.

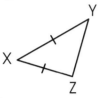

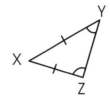

Rule 2
If two angles of a triangle are congruent, then the sides opposite them are also congruent.

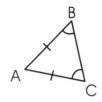

Rule 3
If a triangle is equilateral, then it is also equiangular.

Rule 4
If a triangle is equiangular, then it is also equilateral.

Determine which rule from the box above is used.

1.

2.

3.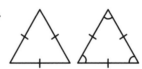

4.

_____   _____   _____   _____

Use the triangles below to answer each question.

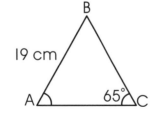

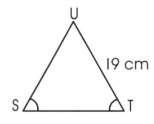

5. m∠BAC = _____

6. $\overline{BC}$ = _____

7. $\overline{SU}$ = _____

8. m∠ABC = m∠ _____

9. m∠STU = _____

10. △STU ≅ _____

Finding Measures of Triangles   Geometric Figures

Find each missing angle or side measure using the information provided in each triangle.

1.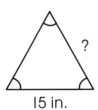
_____

2.
_____

3.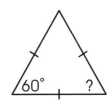
_____

4.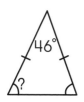
_____

5.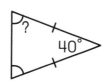
_____

6.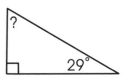
_____

7.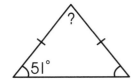
_____

8.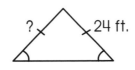
_____

9.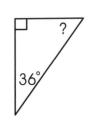
_____

10.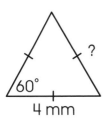
_____

11.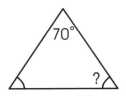
_____

12.
_____

Pythagorean Theorem   Geometric Figures

The **Pythagorean Theorem** is the formula for finding the length of a missing side in a right triangle. In a right triangle, the sum of each leg squared is equal to the length of the hypotenuse squared.

Pythagorean Theorem: $a^2 + b^2 = c^2$ where a and b equal the triangle's legs and c is the hypotenuse.

$c^2 = a^2 + b^2$

$c^2 = 4^2 + 6^2$

$c^2 = 16 + 36$

$c^2 = 52$

$c = \sqrt{52}$

$c = 7.21$ mm

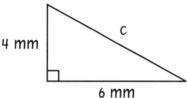

$c^2 = a^2 + b^2$

$a^2 = c^2 - b^2$

$a^2 = 4^2 - 2^2$

$a^2 = 16 - 4$

$a^2 = 12$

$a = \sqrt{12}$

$a = 3.46$ m

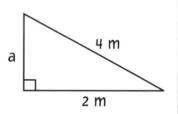

On a separate sheet of paper, use the Pythagorean theorem to find the length of each missing side. Round to the nearest hundredth.

1.
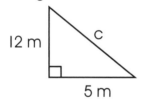

2.
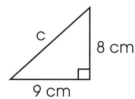

_____

3.
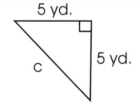

4.
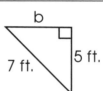

_____

5.
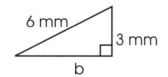

6.
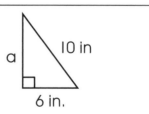

_____

7.

8.

_____

Name: _____ Date: _____

Critical Thinking   Geometric Figures

Answer each question. Show your work.

1. A triangle has sides that measure 4 m, 4 m, and 4 m. What are the measures of its angles? _____ How do you know this?

   _____

   _____

2. Draw and label a triangle congruent to △ABC.

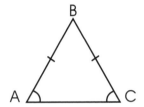

3. A sailboat's sail is a triangle in which two of the angle measures are 90° and 45°. What is the measure of the third angle? _____ How do you know this?

   _____

   _____

   The measures of the triangle's legs are 8 feet and 8 feet. What is the length of the hypotenuse? Round to the nearest tenth. _____

Use the triangle to the right to answer each question.

4. Is ∠YXZ ≅ ∠YZX? _____

   m∠YZX = _____

   m∠XYZ = _____

   $\overline{XY}$ = _____

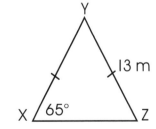

Name: _____ Date: _____

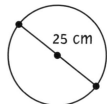

Radius of a Circle
Geometric Figures

A **radius** is a line segment that runs from the center of a circle to a point on the circle. A **diameter** is a line segment that joins two points on a circle and passes through the center. The point in a circle that is equidistant from all points on the circle is the center.

$r = \dfrac{d}{2}$, where r = radius and d = diameter

$r = \dfrac{25}{2}$

$r = 12.5$ cm

25 cm

Find the radius of each circle.

1.
15 cm

_____

2.
6 yd.

_____

3.
13 ft.

_____

4.
2 m

_____

5.
45 in.

_____

6.
21 ft.

_____

7.
4 in.

_____

8.
11 cm

_____

9.
36 cm

_____

10.
112 m

_____

11.
97 yd.

_____

12.
17 ft.

_____

Name: _____ Date: _____

Diameter of a Circle   Geometric Figures

A **diameter** is a line segment that joins two points on a circle and passes through the center.

d = 2r, where d = diameter and r = radius

d = 2 · 15

d = 30 m

15 m

Find the diameter of each circle.

1.

26 m

2.

62 ft.

3.

36 yd.

4.

64 mm

_____   _____   _____   _____

5.

13 in.

6.

33 cm

7.

19 mm

8.

20 in.

_____   _____   _____   _____

9.

7 ft.

10.

46 yd.

11.

22 ft.

12.

67 mm

_____   _____   _____   _____

47

Chords, Tangents, and Secants   Geometric Figures

Chord—a line segment that joins two points on a circle; $\overline{RS}$ is a chord.
The diameter of a circle is its longest chord.
Tangent—a line that intersects a circle at one point;
$\overleftrightarrow{PQ}$ is a tangent.
Secant—a line that intersects a circle at two points;
$\overleftrightarrow{RS}$ is a secant.

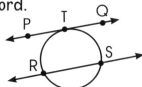

Use the figure to the right to answer each question.

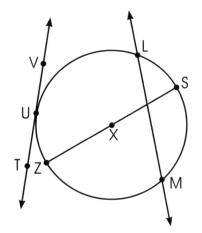

1. Name the shortest chord. _____

2. Name the tangent. _____

3. Name the secant. _____

4. Is $\overline{ZS}$ a secant? _____

   Why or why not? _____

   _____

5. Is $\overline{XS}$ a tangent? _____

   Why or why not? _____

   _____

Use the figure to the right to answer each question.

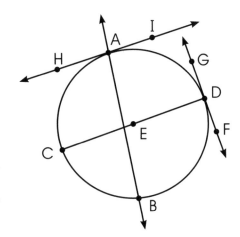

6. Name the longest chord _____

7. Name the tangents. _____ and _____

8. Name the secant. _____

9. Is $\overline{CD}$ a secant? _____

   Why or why not? _____

   _____

10. Is $\overline{ED}$ a chord? _____

    Why or why not? _____

    _____

Identifying Arcs

Geometric Figures

Arc—a continuous portion of a circle's circumference; minor arcs are named with the two points that make the arc; major arcs are named with the three points that make the arc.

Central Angle—an angle with a vertex in the center of a circle; ∠BDA is a central angle.

Minor Arc—an arc created by a central angle with a measure less than or equal to 180°; $\overset{\frown}{BA}$ is a minor arc.

Major Arc—an arc created by a central angle with a measure greater than 180°. $\overset{\frown}{BCA}$ is a major arc.

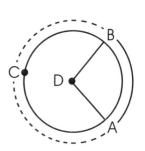

Use the figure to the right to identify each angle or arc by writing *central angle*, *minor arc*, or *major arc*.

1. ∠GIH _____

2. $\overset{\frown}{EFG}$ _____

3. $\overset{\frown}{EG}$ _____

4. $\overset{\frown}{GH}$ _____

5. $\overset{\frown}{FGE}$ _____

6. $\overset{\frown}{HF}$ _____

7. ∠HIF _____

8. ∠EIG _____

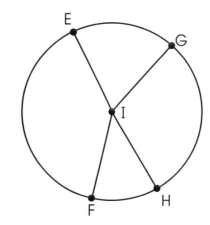

Use the figure to the right to identify each angle or arc by writing central angle, minor arc, or major arc.

9. ∠XZQ _____

10. $\overset{\frown}{XR}$ _____

11. $\overset{\frown}{XPY}$ _____

12. $\overset{\frown}{QR}$ _____

13. $\overset{\frown}{RPQ}$ _____

14. $\overset{\frown}{RP}$ _____

15. $\overset{\frown}{RYQ}$ _____

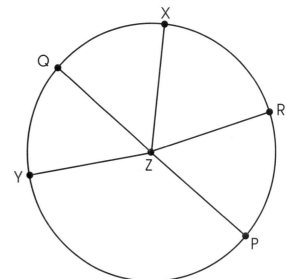

Inscribed Angles

Geometric Figures

Inscribed Angle—an angle formed by two intersecting chords with a vertex on the circle; ∠LMn is an inscribed angle.

Intercepted Arc—an arc of a circle that lies across from a central angle or an inscribed angle. $\overset{\frown}{Ln}$ is an intercepted arc.

Rule: Measures of Central Angles
If an arc is intercepted, then its measure is equal to the measure of the central angle across from it. m∠LOn = 92°

Rule: Measures of Inscribed Angles
If an angle is inscribed in a circle, then its measure is equal to half the measure of its intercepted arc.

$$m\angle LMn = \frac{1}{2}m\overset{\frown}{Ln}$$

$$m\angle LMn = \frac{1}{2} \cdot 92°$$

$$m\angle LMn = 46°$$

Find the measure of each arc or angle in each circle.

1.

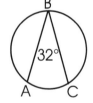

$\overset{\frown}{AC}$ = _____

2.

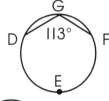

$\overset{\frown}{DEF}$ = _____

3.

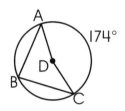

∠ADC = _____

4.

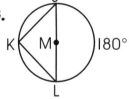

∠JML = _____

5.

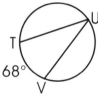

∠TUV = _____

6.

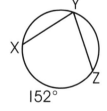

$\overset{\frown}{SU}$ = _____

7.

∠XYZ = _____

8.

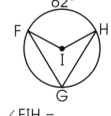

∠FIH = _____

9.

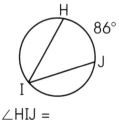

∠HIJ = _____

10.

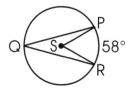

∠PSR = _____

11.

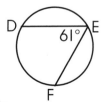

$\overset{\frown}{DF}$ = _____

12.

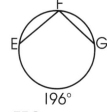

∠EFG = _____

Geometry

Crossword Puzzle

Geometric Figures

Complete the crossword puzzle below.

Across

2. a line that intersects a circle at two points

5. the point in a circle that is equidistant from all points on the circle

7. a line segment that joins two points on a circle and passes through the center

10. an angle with its vertex in the center of a circle

Down

1. an angle formed by two intersecting chords with its vertex on the circle

3. a line segment that joins two points on a circle

4. the part of a circle with a measure less than or equal to 180°

6. a line that intersects a circle at one point

8. a line segment that runs from the center of a circle to a point on the circle

9. an arc created by a central angle with a measure greater than 180°

Name: _____ Date: _____

Find the radius or diameter of each circle.

1.

r = _____

2.

r = _____

3.

d = _____

4.

d = _____

Use the diagram to the right to name each part of the circle.

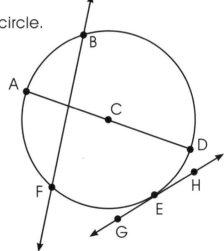

5. diameter _____

6. radius _____

7. shortest chord _____

8. secant _____

9. tangent _____

Draw each part listed below in the circle to the right.

10. Draw chord XY.

11. Draw the diameter XZ.

12. Draw the center point W.

13. Draw a radius WV.

14. Draw secant ST intersecting at V and Z.

15. Draw central angle XWY.

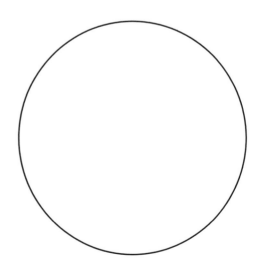

Review   Geometric Figures

Use the diagram to the right to identify each item by writing *diameter, radius, chord, minor arc, major arc, secant,* or *tangent.*

1. $\overline{OQ}$ _____

2. $\overleftrightarrow{ST}$ _____

3. $\overset{\frown}{QL}$ _____

4. $\overleftrightarrow{ON}$ _____

5. $\overset{\frown}{NMO}$ _____

6. $\overline{LM}$ _____

7. $\overset{\frown}{NP}$ _____

8. $\overline{QR}$ _____

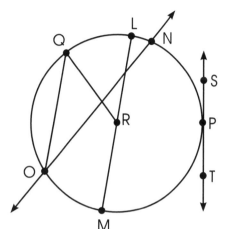

Find the measure of each arc or angle.

9.

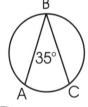

$\overset{\frown}{AC}$ = _____

10.

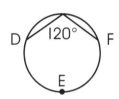

$\overset{\frown}{DEF}$ = _____

11.

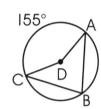

∠ADC = _____

12.

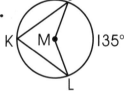

∠JML = _____

13.

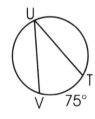

∠TUV = _____

14.

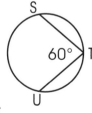

$\overset{\frown}{SU}$ = _____

15.

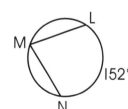

∠LMN = _____

16.

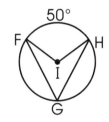

∠FIH = _____

Name: _____   Date: _____

Area and Perimeter   Area, Perimeter, and Volume

Area—the measure of space within a figure
Perimeter—the measure of distance around a polygon

Calculating Area and Perimeter of a Square

Area = s^2 or $s \cdot s$, where s = side length

Perimeter = 4s, or $s + s + s + s$, where
s = side length

$A = s^2$	$P = 4s$	
$A = 5^2$	$P = 4 \cdot 5$	
$A = 25$ m²	$P = 20$ m	5 m

Calculating Area and Perimeter of a Rectangle

Area = lw, where l = length and w = width

Perimeter = $2l + 2w$ (or $l + w + l + w$,
where l = length and w = width)

$A = lw$ $P = 2l + 2w$ 6 yd.

$A = 2 \cdot 6$ $P = (2 \cdot 2) + (2 \cdot 6)$

$A = 12$ yd.² $P = 4 + 12$

                        $P = 16$ yd.   2 yd.

Find the area and perimeter of each figure. Show your work on a separate sheet of paper.

1.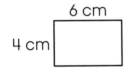
6 cm
4 cm

A = _____   P = _____

2.
12 in.

A = _____   P = _____

3.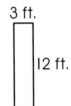
3 ft.
12 ft.

A = _____   P = _____

4.
8 yd.
2 yd.

A = _____   P = _____

5.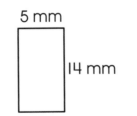
5 mm
14 mm

A = _____   P = _____

6.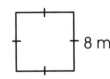
8 m

A = _____   P = _____

7.
4 in.

A = _____   P = _____

8.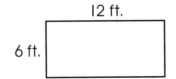
12 ft.
6 ft.

A = _____   P = _____

9.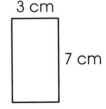
3 cm
7 cm

A = _____   P = _____

Area and Perimeter

Area, Perimeter, and Volume

Calculating Area and Perimeter of a Parallelogram

Area = bh, where b = base and h = height
(height must form a right angle with base)

Perimeter = 2a + 2b, where a = one side and b = other side

A = bh	P = 2a + 2b
A = 2 · 5	P = (2 · 6) + (2 · 2)
A = 10 in.²	P = 12 + 4
	P = 16 in.

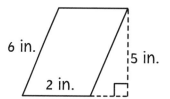

Find the area and perimeter of each parallelogram. Show your work on a separate sheet of paper.

1.

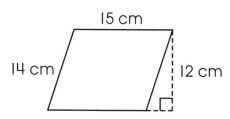

A = _____ P = _____

2.

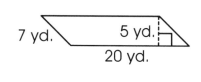

A = _____ P = _____

3.

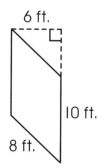

A = _____ P = _____

4.

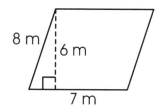

A = _____ P = _____

5.

A = _____ P = _____

6.

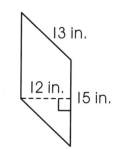

A = _____ P = _____

7.

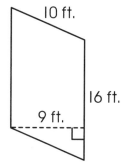

A = _____ P = _____

8.

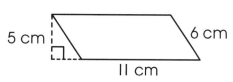

A = _____ P = _____

9.

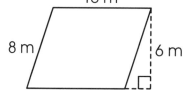

A = _____ P = _____

Area and Perimeter

Area, Perimeter, and Volume

Calculating Area and Perimeter of a Rhombus
A **rhombus** is a parallelogram with four congruent sides.

Area $= \frac{1}{2}(d_1 \cdot d_2)$, where d_1 and d_2 are the diagonals

Perimeter $= 4a$, where $a =$ side length

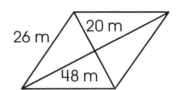

$A = \frac{1}{2}(d_1 \cdot d_2)$ $P = 4a$

$A = \frac{1}{2}(20 \cdot 48)$ $P = 4 \cdot 26$

$A = \frac{1}{2}(960)$ $P = 104$ m

$A = 480$ m²

Use a calculator to find the area and perimeter of each rhombus. Show your work on a separate sheet of paper.

1.

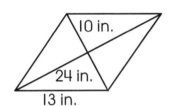

A = _____ P = _____

2.

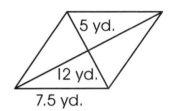

A = _____ P = _____

3.

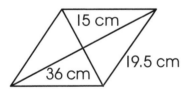

A = _____ P = _____

4.

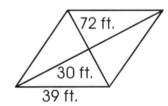

A = _____ P = _____

5.

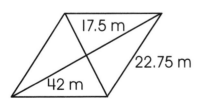

A = _____ P = _____

6.

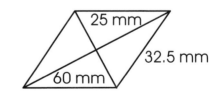

A = _____ P = _____

Area and Perimeter
Calculating Area and Perimeter of a Triangle

Area, Perimeter, and Volume

Area $= \frac{1}{2}$ bh, where b = base and h = height (height must form a right angle with the base)

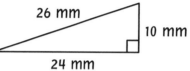

height
base

Perimeter $= a + b + c$, where a, b, and c are side lengths

$A = \frac{1}{2}$ bh $P = a + b + c$

$A = \frac{1}{2} \cdot 24 \cdot 10$ $P = 10 + 24 + 26$

$A = 120$ mm² $P = 60$ mm

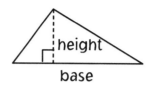

26 mm
10 mm
24 mm

Use a calculator to find the area and perimeter of each triangle. Show your work on a separate sheet of paper.

1.

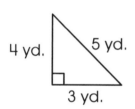

4 yd.   5 yd.
3 yd.

A = _____ P = _____

2.

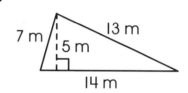

7 m   13 m
5 m
14 m

A = _____ P = _____

3.

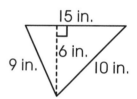

15 in.
6 in.
9 in.   10 in.

A = _____ P = _____

4.

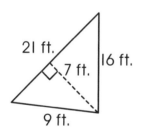

21 ft.
7 ft.   16 ft.
9 ft.

A = _____ P = _____

5.

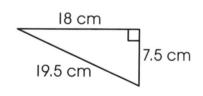

18 cm
7.5 cm
19.5 cm

A = _____ P = _____

6.

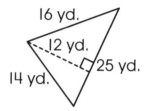

16 yd.
12 yd.
14 yd.   25 yd.

A = _____ P = _____

7.

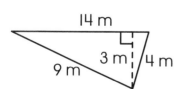

14 m
3 m   4 m
9 m

A = _____ P = _____

8.

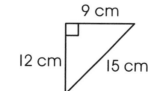

9 cm
12 cm   15 cm

A = _____ P = _____

9.

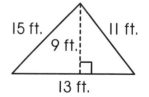

15 ft.   11 ft.
9 ft.
13 ft.

A = _____ P = _____

Name: _____ Date: _____

Area and Perimeter

Calculating Area and Perimeter of a Trapezoid

Area = $\frac{1}{2}$ h(b$_1$ + b$_2$), where b$_1$ and b$_2$ represent each base and h = height (height must form a right angle with a base)

Perimeter = a + b + c + d, where a, b, c, and d are side lengths

Area, Perimeter, and Volume

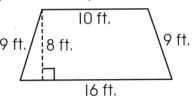

$A = \frac{1}{2} h(b_1 + b_2)$ $P = a + b + c + d$

$A = \frac{1}{2} \cdot 8(10 + 16)$ $P = 10 + 16 + 9 + 9$

$A = \frac{1}{2} \cdot 8(26)$ $P = 44$ ft.

$A = 4(26)$

$A = 104$ ft.^{2}

Use a calculator to find the area and perimeter of each trapezoid. Show your work on a separate sheet of paper.

1.

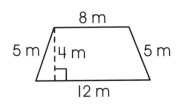

A = _____ P = _____

2.

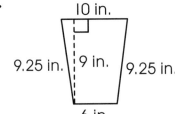

A = _____ P = _____

3.

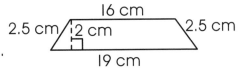

A = _____ P = _____

4.

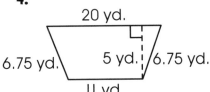

A = _____ P = _____

5.

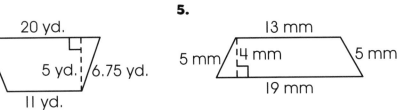

A = _____ P = _____

6.

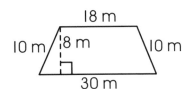

A = _____ P = _____

7. Calculate the area of the trapezoid below without using the formula for the area of a trapezoid. Use the formula to check your answer.

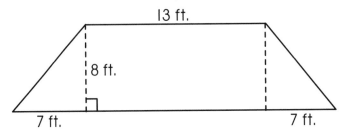

Name: _____ Date: _____

Area and Circumference

Area, Perimeter, and Volume

Calculating Area and Circumference of a Circle

Circumference—the perimeter of a circle

Pi (π)—the ratio of a circle's circumference to its diameter

Area = πr^2, where r = radius and π = 3.14

Circumference = πd or $2\pi r$, where
π = 3.14, d = diameter, and r = radius

$A = \pi r^2$ $C = \pi d$

$A = 3.14 \cdot 6^2$ $C = 3.14 \cdot 12$

$A = 3.14 \cdot 36$ $C = 37.68$ mm

$A = 113.04$ mm^2

Use a calculator to find the area and circumference of each circle. Show your work on a separate sheet of paper. Round to the nearest hundredth.

1.

A = _____ C = _____

2.

A = _____ C = _____

3.

A = _____ C = _____

4.

A = _____ C = _____

5.

A = _____ C = _____

6.

A = _____ C = _____

 Geometry

Name: _____   Date: _____

Area and Perimeter Practice   Area, Perimeter, and Volume

Use a calculator to find the area and perimeter of each polygon. Show your work on a separate sheet of paper.

1.

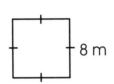

A = _____ P = _____

2.

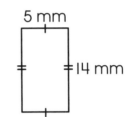

A = _____ P = _____

3.

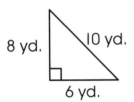

A = _____ P = _____

4.

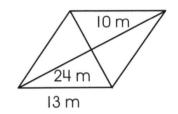

A = _____ P = _____

5.

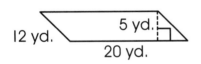

A = _____ P = _____

6.

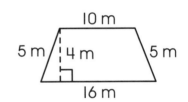

A = _____ P = _____

7.

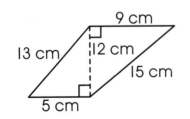

A = _____ P = _____

8.

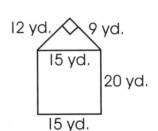

A = _____ P = _____

Critical Thinking — Area, Perimeter, and Volume

Answer each question. Show your work on a separate sheet of paper. Hint: Draw pictures to help.

1. Sara has a job installing carpet. The room she is carpeting is 10 m by 20 m. How much carpet does she need?

2. Juan is planning to build a deck and needs to know how much material to buy. The deck will wrap around two sides of the house. The house measures 54 feet by 60 feet. He wants the deck to be 10 feet wide. How many square feet of material does Juan need to buy?

3. Jamie is making a sail for a boat. The sail is a right triangle and needs to be 16 feet high. He knows that he needs 96 square feet of sail to support his boat. How long does the base of the sail need to be?

4. Teresa is making a kite in the shape of a rhombus. She has two poles to support the kite. The diagonal poles' lengths are 15 inches and 36 inches. How much material does Teresa need to buy in order to make her kite?

5. A rectangle has an area of 45 in.^{2} and a length of 5 in. Find the width and the perimeter.

6. A square has a perimeter of 100 cm. Find the lengths of the sides and the area.

7. A parallelogram has a base of 17 mm and a height of 3 mm. Find the area.

8. Find the circumference of a circle with a diameter of 24 m.

9. Find the radius of a circle with an area of 50.24 cm^2.

10. A circle has an area of 153.86 in.^{2} and a circumference of 43.96 in. Find the diameter of the circle.

Name: _____ Date: _____

Surface Area

Area, Perimeter, and Volume

Surface Area—the sum of the areas of all of the faces in a three-dimensional figure

Calculating Surface Area of a Cube

$SA = 6B$, where B = area of the base (s^2)

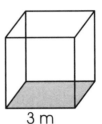

3 m

$B = s^2$	$SA = 6B$
$B = 3^2$	$SA = 6 \cdot 9$
$B = 9\ m^2$	$SA = 54\ m^2$

Use a calculator to find the surface area of each cube. Show your work on a separate sheet of paper.

1.

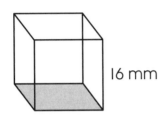

16 mm

SA = _____

2.

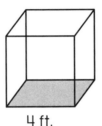

4 ft.

SA = _____

3.

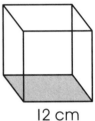

12 cm

SA = _____

4.

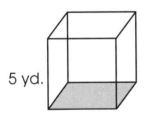

5 yd.

SA = _____

5.

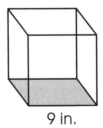

9 in.

SA = _____

6.

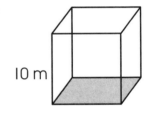

10 m

SA = _____

7.

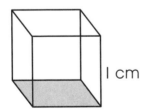

1 cm

SA = _____

8.

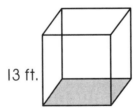

13 ft.

SA = _____

9.

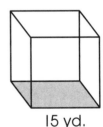

15 yd.

SA = _____

Name: _____ Date: _____

Surface Area

Area, Perimeter, and Volume

Rectangular Prism—a prism that has two rectangular bases and four rectangular faces

Calculating Surface Area of a Rectangular Prism

SA = 2B + Ph, where B = area of the base (lw), P = perimeter of the base (2l + 2w), and h = height

Step 1: Find the base and calculate its area.

B = lw

B = 6 x 5

B = 30 cm²

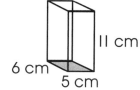

11 cm
6 cm
5 cm

Step 2: Calculate the perimeter of the base.

P = 2(6) + 2(5)

P = 22 cm

Step 3: Find the height of the prism.

h = 11 cm

Step 4: Substitute the height, the base's area, and the base's perimeter into the formula.

SA = 2B + Ph

SA = 2(30) + (22)(11)

SA = 60 + 242

SA = 302 cm²

Use a calculator to find the surface area of each rectangular prism. Show your work on a separate sheet of paper.

1.

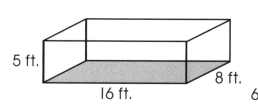

5 ft.
16 ft.
8 ft.

SA = _____

2.

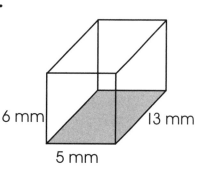

6 mm
5 mm
13 mm

SA = _____

3.

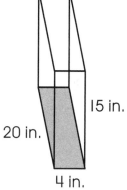

15 in.
20 in.
4 in.

SA = _____

4.

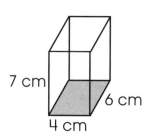

7 cm
6 cm
4 cm

SA = _____

5.

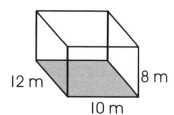

12 m
8 m
10 m

SA = _____

6.

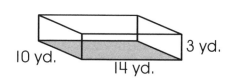

10 yd.
14 yd.
3 yd.

SA = _____

Name: _____   Date: _____

Surface Area
Area, Perimeter, and Volume

Square Pyramid—a pyramid that has a square base and triangular faces that meet at a common vertex

Calculating Surface Area of a Square Pyramid
$SA = B + \frac{1}{2}Pl$, where B = area of the base (s^2),
P = perimeter of the base ($4s$), and l = slant height

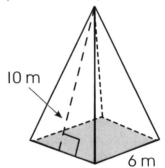

Slant height is the distance from the base to the vertex. On a solid object's face, the height must form a right angle with the edge of the base.

Step 1: Find the area of the base.

$$B = s^2$$
$$B = 6^2$$
$$B = 36 \text{ m}^2$$

Step 2: Find the perimeter of the base.

$$P = 4s$$
$$P = 4(6)$$
$$P = 24 \text{ m}$$

Step 3: Find the pyramid's slant height.

$$l = 10 \text{ m}$$

Step 4: Substitute the base's area, the base's perimeter, and the slant height into the formula.

$$SA = B + \frac{1}{2}Pl$$
$$SA = 36 + \frac{1}{2}(24)(10)$$
$$SA = 36 + \frac{1}{2}(240)$$
$$SA = 36 + 120$$
$$SA = 156 \text{ m}^2$$

Use a calculator to find the surface area of each square pyramid. Show your work on a separate sheet of paper.

1.

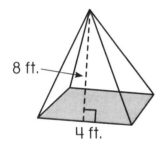

8 ft.

4 ft.

SA = _____

2.

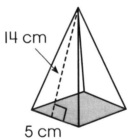

14 cm

5 cm

SA = _____

3.

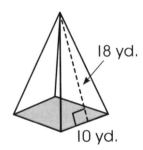

18 yd.

10 yd.

SA = _____

4.

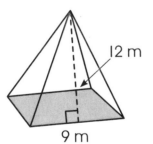

12 m

9 m

SA = _____

5.

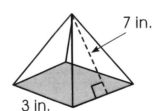

7 in.

3 in.

SA = _____

6.

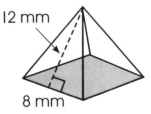

12 mm

8 mm

SA = _____

Geometry

Surface Area
Area, Perimeter, and Volume

Calculating Surface Area of a Cylinder

$SA = 2B + Ch$, where B = area of the base (πr^2),
C = circumference of the base (πd), and h = height

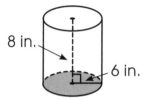

8 in.
6 in.

Step 1: Find the base and calculate its area.

$B = \pi r^2$

$B = 3.14 \cdot 6^2$

$B = 3.14 \cdot 36$

$B = 113.04$ in.^{2}

Step 2: Calculate the circumference of the base.

$C = \pi d$

$C = 3.14 \cdot 12$

$C = 37.68$ in.

Step 3: Find the cylinder's height.

$h = 8$ in.

Step 4: Substitute the base's area, the base's circumference, and the height into the formula.

$SA = 2B + Ch$

$SA = 2(113.04) + (37.68)(8)$

$SA = 226.08 + 301.44$

$SA = 527.52$ in.^{2}

Use a calculator to find the surface area of each cylinder. Show your work on a separate sheet of paper. Round to the nearest hundredth.

1.

3 mm

7 mm

SA = _____

2.

18 ft.

12 ft.

SA = _____

3.

3 cm

9 cm

SA = _____

4.

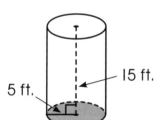

15 ft.

5 ft.

SA = _____

5.

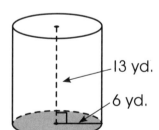

13 yd.

6 yd.

SA = _____

6.

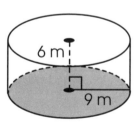

6 m

9 m

SA = _____

Name: _____   Date: _____

Surface Area

Calculating Surface Area of a Cone

$SA = B + \frac{1}{2}Cl$, where B = area of the base (πr^2),
C = circumference of the base (πd), and
l = slant height

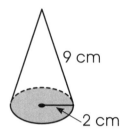

9 cm

2 cm

Step 1: Find the base and calculate its area.

$B = \pi r^2$

$B = 3.14 \times 2^2$

$B = 3.14 \times 4$

$B = 12.56 \text{ cm}^2$

Step 2: Calculate the circumference of the base.

$C = \pi d$

$C = 3.14 \times 4$

$C = 12.56 \text{ cm}$

Step 3: Find the cylinder's slant height.

$l = 9 \text{ cm}$

Step 4: Substitute the slant height, the base's area, and the base's circumference into the formula.

$SA = B + \frac{1}{2}Cl$

$SA = 12.56 + \frac{1}{2}(12.56)(9)$

$SA = 12.56 + \frac{1}{2}(113.04)$

$SA = 12.56 + 56.52$

$SA = 69.08 \text{ cm}^2$

Use a calculator to find the surface area of each cone. Show your work on a separate sheet of paper. Round to the nearest hundredth.

1.

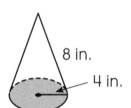

8 in.

4 in.

SA = _____

2.

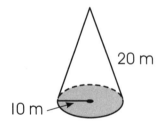

20 m

10 m

SA = _____

3.

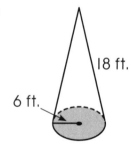

18 ft.

6 ft.

SA = _____

4.

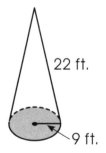

22 ft.

9 ft.

SA = _____

5.

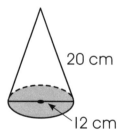

20 cm

12 cm

SA = _____

6.

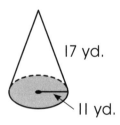

17 yd.

11 yd.

SA = _____

Name: _____ Date: _____

Surface Area
Calculating Surface Area of a Sphere

SA = 4πr²
SA = 4(3.14)(12²)
SA = 1,808.64 mm²

Area, Perimeter, and Volume

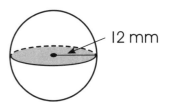
12 mm

Use a calculator to find the surface area of each sphere. Show your work on a separate sheet of paper. Round to the nearest hundredth.

1. 8 ft.

SA = _____

2. 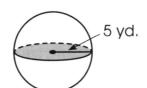 5 yd.

SA = _____

3. 10 in.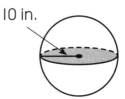

SA = _____

4. 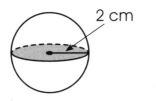 2 cm

SA = _____

5. 9 ft.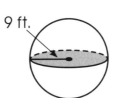

SA = _____

6. 26 mm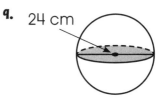

SA = _____

7. 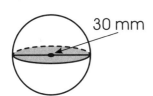 30 mm

SA = _____

8. 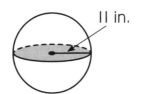 11 in.

SA = _____

9. 24 cm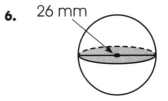

SA = _____

10. 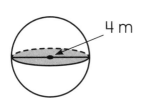 4 m

SA = _____

11. 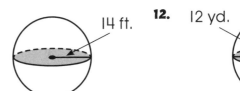 14 ft.

SA = _____

12. 12 yd.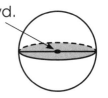

SA = _____

 Geometry

Name: _____ Date: _____

Surface Area Practice   Area, Perimeter, and Volume

Use a calculator to find the surface area of each rectangular prism. Show your work on a separate sheet of paper. Remember: (SA = 2B + Ph)

1.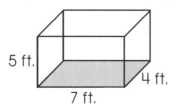

5 ft.
4 ft.
7 ft.

SA = _____

2.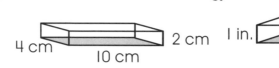

4 cm
10 cm
2 cm

SA = _____

3.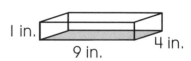

1 in.
9 in.
4 in.

SA = _____

Use a calculator to find the surface area of each square pyramid. Show your work on a separate sheet of paper. Remember: (SA = B + $\frac{1}{2}$Pl)

4.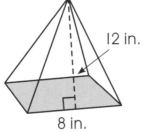

12 in.
8 in.

SA = _____

5.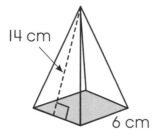

14 cm
6 cm

SA = _____

6.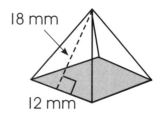

18 mm
12 mm

SA = _____

Use a calculator to find the surface area of each cylinder. Show your work on a separate sheet of paper. Remember: (SA = 2B + Ch)

7.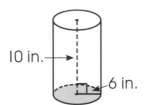

10 in.
6 in.

SA = _____

8.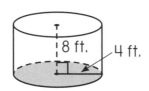

8 ft.
4 ft.

SA = _____

4.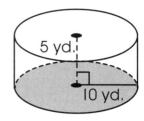

5 yd.
10 yd.

SA = _____

Use a calculator to find the surface area of each cone. Show your work on a separate sheet of paper. Remember: (SA = B + $\frac{1}{2}$Cl)

10.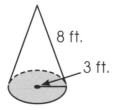

8 ft.
3 ft.

SA = _____

11.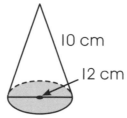

10 cm
12 cm

SA = _____

12.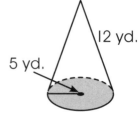

12 yd.
5 yd.

SA = _____

Name: _____ Date: _____

Volume
Area, Perimeter, and Volume

Volume—the amount of space within a three-dimensional figure

Calculating Volume of a Cube
Volume = a^3, where a = side length

$V = a^3$
$V = 8^3$
$V = 512 \text{ m}^3$

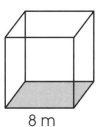

8 m

Use a calculator to find the volume of each cube. Show your work on a separate sheet of paper.

1.
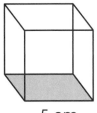
5 cm

V = _____

2.
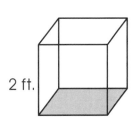
2 ft.

V = _____

3.
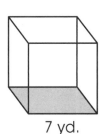
7 yd.

V = _____

4.
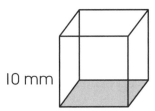
10 mm

V = _____

5.
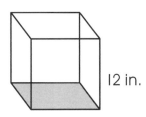
12 in.

V = _____

6.
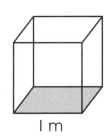
1 m

V = _____

7.
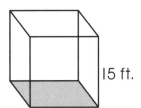
15 ft.

V = _____

8.
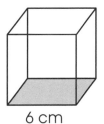
6 cm

V = _____

9.
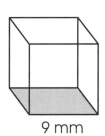
9 mm

V = _____

Name: _____ Date: _____

Volume

Area, Perimeter, and Volume

Calculating Volume of a Rectangular Prism

$V = Bh$, where B = area of the base (lw)
and h = height

Step 1: Find the base and calculate its area.

$B = lw$

$B = 12 \cdot 8$

$B = 96$ cm²

Step 2: Find the height.

$h = 4$ cm

Step 2: Substitute the area of the base and height into the formula.

$V = Bh$

$V = 96 \cdot 4$

$V = 384$ cm³

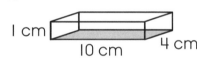

Use a calculator to find the volume of each rectangular prism. Show your work on a separate sheet of paper.

1.

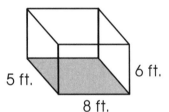

V = _____

2.

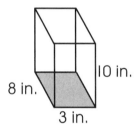

V = _____

3.

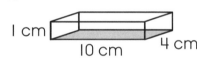

V = _____

4.

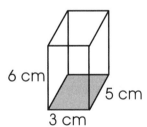

V = _____

5.

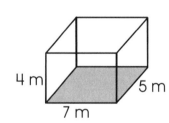

V = _____

6.

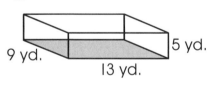

V = _____

7.

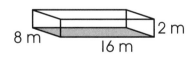

V = _____

8.

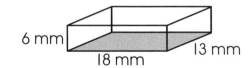

V = _____

9.

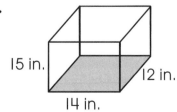

V = _____

Name: _____   Date: _____

Volume
Area, Perimeter, and Volume

Calculating Volume of Square Pyramids and Cones
$V = \frac{1}{3}Bh$, where B = area of the base and h = height (forming a right angle to the base)
Remember: (square: $A = s^2$; triangle: $A = \frac{1}{2}bh$; circle: $A = \pi r^2$)

Calculating Volume of a Square Pyramid
Step 1: Find the base and calculate its area.

$B = s^2$
$B = 5^2$
$B = 25$ ft.^{2}

Step 2: Find the height of the pyramid.

$h = 11$ ft.

Step 3: Substitute the base's area and height into the formula.

$V = \frac{1}{3}Bh$
$V = \frac{1 \cdot (25 \cdot 11)}{3}$
$V = \frac{275}{3}$
$V = 91.67$ ft.^{3}

5 ft.

11 ft.

Calculating Volume of a Cone
Step 1: Find the base and calculate its area.

$B = \pi r^2$
$B = 3.14 \cdot 3^2$
$B = 3.14 \cdot 9$
$B = 28.26$ m^2

Step 2: Find the height of the cone.

$h = 6$ m

Step 3: Substitute the height and area of base into the formula.

$V = \frac{1}{3}Bh$
$V = \frac{1 \cdot (28.26 \cdot 6)}{3}$
$V = \frac{169.56}{3}$
$V = 56.52$ m^3

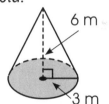

6 m
3 m

Use a calculator to find the volume of each square pyramid. Show your work on a separate sheet of paper. Round to the nearest hundredth.

1.
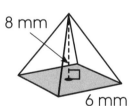
8 mm
6 mm

V = _____

2.
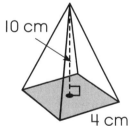
10 cm
4 cm

V = _____

3.
13 in.
8 in.

V = _____

Use a calculator to find the volume of each cone. Show your work on a separate sheet of paper. Round to the nearest hundredth.

4.
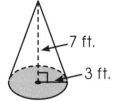
7 ft.
3 ft.

V = _____

5.
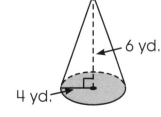
6 yd.
4 yd.

V = _____

6.
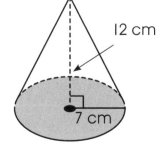
12 cm
7 cm

V = _____

Volume

Area, Perimeter, and Volume

Calculating Volume of a Cylinder

$V = Bh$, where B = area of the base (πr^2) and h = height

Step 1: Find the base and calculate its area.

$B = \pi r^2$

$B = 3.14 \cdot 5^2$

$B = 3.14 \cdot 25$

$B = 78.5 \text{ cm}^2$

Step 2: Find the height of the cylinder.

$h = 12 \text{ cm}$

Step 2: Substitute the height and area of the base into the formula.

$V = Bh$

$V = 78.5 \cdot 12$

$V = 942 \text{ cm}^3$

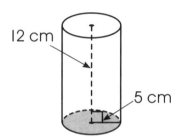
12 cm
5 cm

Use a calculator to find the volume of each cylinder. Show your work on a separate sheet of paper. Round to the nearest hundredth.

1.

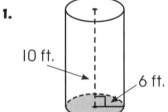
10 ft.
6 ft.

V = _____

2.

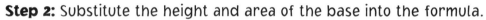
9 in.
3 in.

V = _____

3.

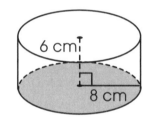
6 cm
8 cm

V = _____

4.

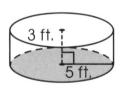
3 ft.
5 ft.

V = _____

5.

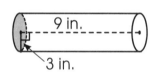
2 m
11 m

V = _____

6.

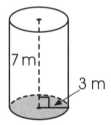
7 m
3 m

V = _____

7.

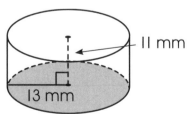
11 mm
13 mm

V = _____

8.

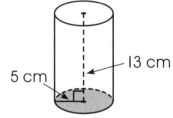
5 cm
13 cm

V = _____

9.

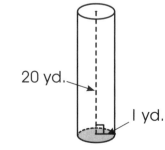
20 yd.
1 yd.

V = _____

Volume

Area, Perimeter, and Volume

Calculating Volume of a Sphere

$V = \frac{4}{3}\pi r^3$

$\quad V = \frac{4}{3} \cdot 3.14 \cdot 6^3$

$\quad V = \frac{4 \cdot 3.14 \cdot 216}{3}$

$\quad V = \frac{2,712.96}{3}$

$\quad V = 904.32 \text{ in.}^3$

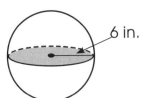

6 in.

Use a calculator to find the volume of each sphere. Show your work on a separate sheet of paper. Round to the nearest hundredth.

1.

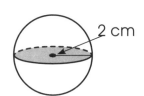

2 cm

V = _____

2.

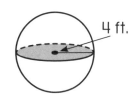

4 ft.

V = _____

3.

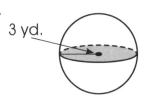

3 yd.

V = _____

4.

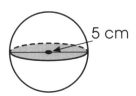

5 cm

V = _____

5.

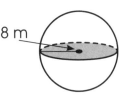

8 m

V = _____

6.

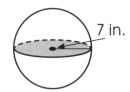

7 in.

V = _____

7.

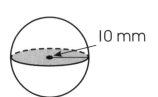

10 mm

V = _____

8.

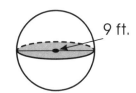

9 ft.

V = _____

9.

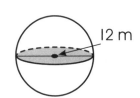

12 m

V = _____

Name: _____ Date: _____

Critical Thinking

Area, Perimeter, and Volume

Use a calculator to answer each question. Show your work on a separate sheet of paper. Round to the nearest hundredth. Hint: Draw pictures to help.

1. Monty is building a wooden frame for a garden with a friend. He wants it to measure 10 feet long, 5 feet wide, and 1 foot high. What will the volume of the frame be?

2. Karen visited an aquarium with a cylinder-shaped fish tank. She saw a sign that said that the tank is 45 feet high and 20 feet in diameter. What is the surface area of the fish tank?

3. Your neighbors are having concrete poured for their driveway's foundation. The foundation will be 0.25 feet deep and 24 feet long by 12 feet wide. How much concrete will be needed to complete the job?

4. Lamar is taking a beach ball with him to the beach. What is the volume of the beach ball if its radius is 18 inches?

5. Find the volume of a cone with a height of 16 cm and a radius of 4 cm.

6. A square pyramid has a height of 12 m and a base length of 5 m. What is the volume of the square pyramid?

7. Find the surface area of a cube with a side length of 8 mm.

8. Find the volume of a cube with a side length of 4 yards.

9. Find the surface area of a cylinder with a height of 5 cm and a radius of 5 cm.

10. Find the surface area of a square pyramid with a base length of 10 m and a slant height of 12 m.

Review   Area, Perimeter, and Volume

Use a calculator to find the area and the perimeter or circumference of each figure.
Show your work on a separate sheet of paper.

1.
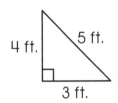
A = _____ P = _____

2.
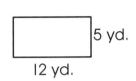
A = _____ P = _____

3.
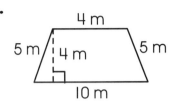
A = _____ P = _____

4.

A = _____ C = _____

5.

A = _____ C = _____

6.
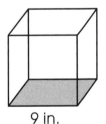
A = _____ P = _____

Use a calculator to find the surface area of each solid. Show your work on a separate
sheet of paper.

7.
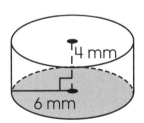
SA = _____

8.
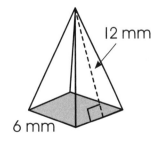
SA = _____

9.
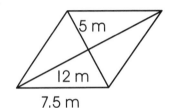
SA = _____

10.
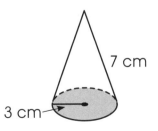
SA = _____

11.
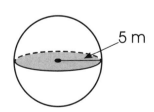
SA = _____

12.
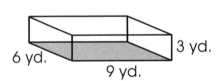
SA = _____

Name: _____ Date: _____

Review   Area, Perimeter, and Volume

Use a calculator to find the volume of each solid. Show your work on a separate sheet of paper.

1.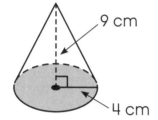

9 cm

4 cm

V = _____

2.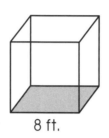

8 ft.

V = _____

3.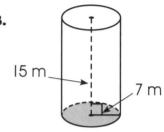

15 m

7 m

V = _____

4.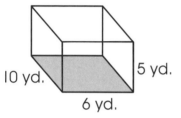

10 yd.

5 yd.

6 yd.

V = _____

5.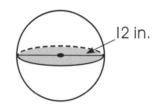

12 in.

V = _____

6.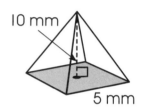

10 mm

5 mm

V = _____

7.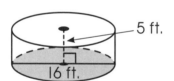

5 ft.

16 ft.

V = _____

8.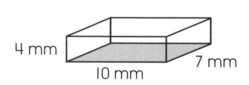

4 mm

10 mm

7 mm

V = _____

9.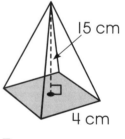

15 cm

4 cm

V = _____

10.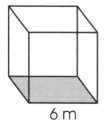

6 m

V = _____

11.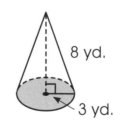

8 yd.

3 yd.

V = _____

12.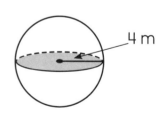

4 m

V = _____

Transformations

Transformations and Symmetry

A **transformation** is a movement of specific change to a geometric figure. Reflections, rotations, and translations are transformations that **do not** result in an object changing size or shape. Therefore, the transformed objects are congruent.

Reflection (or flip)—a transformation that produces a mirror image of a figure reflected across a fixed line

Rotation (or turn)—a transformation that turns a figure around a fixed point

Translation (or slide)—a transformation that shifts or moves a shape in one direction without rotation

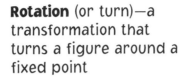
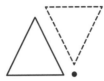
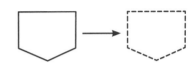

Name the congruent angles of each reflection.

1. ∠RQS ≅ ∠_____ ; ∠QRS ≅ ∠_____

∠QSR ≅ ∠_____

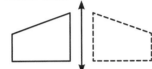

2. ∠_____ ≅ ∠_____ ; ∠_____ ≅ ∠_____

∠_____ ≅ ∠_____ ; ∠_____ ≅ ∠_____

∠_____ ≅ ∠_____

Name the congruent angles of each rotation.

3. ∠_____ ≅ ∠_____ ; ∠_____ ≅ ∠_____

∠_____ ≅ ∠_____

4. ∠_____ ≅ ∠_____ ; ∠_____ ≅ ∠_____

∠_____ ≅ ∠_____ ; ∠_____ ≅ ∠_____

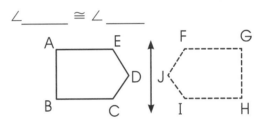

Name the congruent angles of each translation.

5. ∠ABC ≅ ∠_____ ; ∠BCD ≅ ∠_____

∠CDA ≅ ∠_____ ; ∠DAB ≅ ∠_____

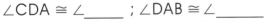
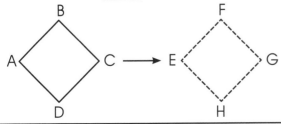

6.

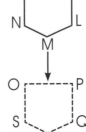

∠_____ ≅ ∠_____

∠_____ ≅ ∠_____

∠_____ ≅ ∠_____

∠_____ ≅ ∠_____

∠_____ ≅ ∠_____

Name: _____   Date: _____

Identifying Transformations

Circle the name of the correct transformation for each pair of figures or shapes below. If none of the transformations have been performed, circle *none*.

1. 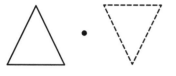   reflection   rotation   translation   none

2. 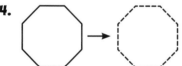   reflection   rotation   translation   none

3.    reflection   rotation   translation   none

4. 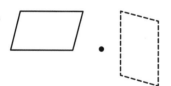   reflection   rotation   translation   none

5.   reflection   rotation   translation   none

6. 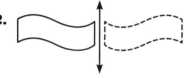   reflection   rotation   translation   none

7.    reflection   rotation   translation   none

8.    reflection   rotation   translation   none

Name: _____   Date: _____

Drawing Reflections
Transformations and Symmetry

Step 1: Measure the distance between each vertex and the line of reflection and draw new vertices equidistant from the line on the opposite side.

Step 2: Draw the new flipped image by connecting the vertices you have marked.

Step 3: Check your work by measuring each side and angle to make sure the two figures are congruent.

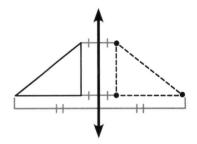

Draw the reflection of each figure.

1.

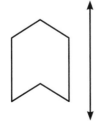

2.

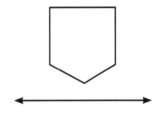

3.

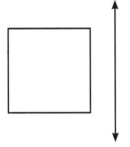

4.

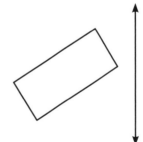

5.

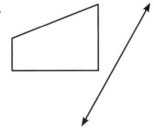

6.

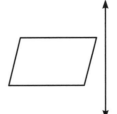

Drawing Translations Transformations and Symmetry

To draw a translation, move the figure in the direction of the arrow to a new place.

Draw the translation of each shape.

1.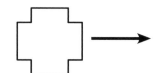

2.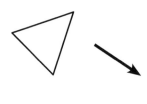

3.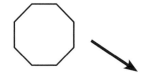

4.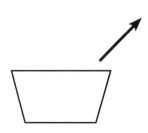

5.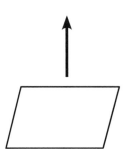

6.

Name: _____ Date: _____

Understanding Degrees

Transformations and Symmetry

Understanding degree measurements will help determine how to rotate a figure.

Measuring 360° clockwise

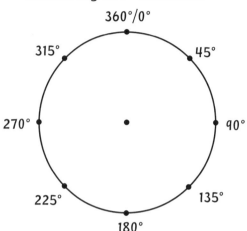

Measuring 360° counterclockwise

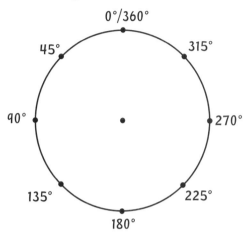

To find the measurement of an unmarked degree, subtract the given measure from 360°. Always measure in a clockwise direction unless counterclockwise is specified.

Use the circles above to answer each question.

1. 90° measured in a clockwise direction is the same as _____ measured in a counterclockwise direction.

2. 225° measured in a counterclockwise direction is the same as _____ measured in a clockwise direction.

3. 315° measured in a clockwise direction is the same as _____ measured in a counterclockwise direction.

4. 360° measured in a clockwise direction is the same as _____ measured in a counterclockwise direction.

5. 25° measured in a counterclockwise direction is the same as _____ measured in a clockwise direction.

6. 132° measured in a counterclockwise direction is the same as _____ measured in a clockwise direction.

7. 50° measured in a clockwise direction is the same as _____ measured in a counterclockwise direction.

8. 39° measured in a clockwise direction is the same as _____ measured in a counterclockwise direction.

Drawing Rotations
Transformations and Symmetry

Rotating a Shape around a Point

Imagine nailing the center of a triangle and spinning it. Depending on where you stop rotating the triangle, you will likely get a different image. Therefore, stating a specific degree that the shape will be rotated is important. The following triangle has been rotated 45°, 90°, 180°, and 270° around a point.

0°   45°   90°   180°   270°
(or 90° counterclockwise)

Drawing Rotations

To draw a shape rotated 90°:

Step 1: Draw a line segment connecting the point around which the figure will be rotated to a vertex on the figure. Measure the length of the segment.

Step 2: Measure a 90° angle by using a protractor and draw a line from the point of rotation to form a 90° angle.

Step 3: On the line created in Step 2, place a new point to form a line segment that is the same length as the segment between the figure's vertex and the point of rotation. (See Step 1.) This becomes the new corresponding vertex of the rotated figure.

Step 4: Follow the same steps for each vertex.

Step 5: Connect the vertices.

(Substitute the correct degree measurement for other rotations.)

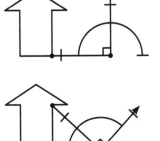
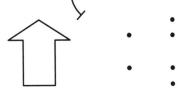
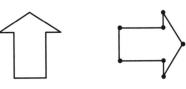

Use a protractor and a ruler to draw each rotation.

1. 180° rotation

2. 135° rotation

3. 45° counterclockwise rotation

4. 90° counterclockwise rotation

Drawing More Rotations

Transformations and Symmetry

Use a protractor and a ruler to draw each rotation.

1. 45° counterclockwise rotation

2. 120° rotation

3. 90° rotation

4. 45° rotation

5. 90° counterclockwise rotation

6. 150° rotation

7. 60° rotation

8. 135° rotation

Name: _____   Date: _____

Dilations   Transformations and Symmetry

A **dilation**, or scaling, is a transformation that changes the size of a figure without changing its shape. The two figures are **similar** but not congruent.

This dilated figure has changed in size, but not shape.

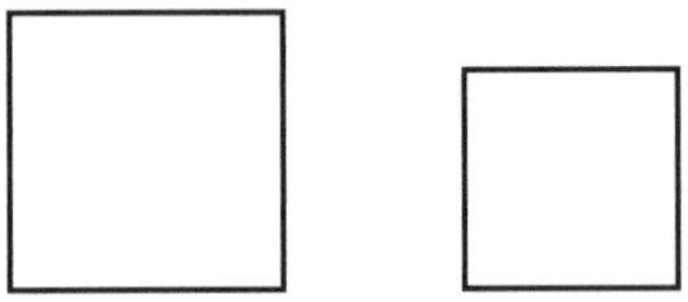

Determine if each figure has been dilated and write *Yes* or *No*. Explain your answers.

	Dilated?	Explain
1. 		
2. 		
3. 		
4. 		
5. 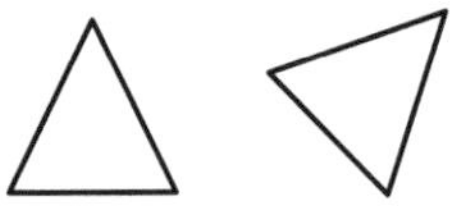		
6. 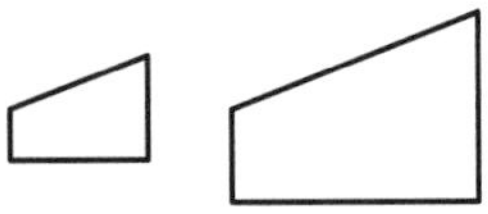		
7. 		
8. 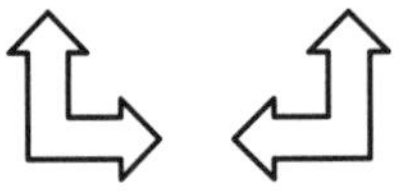		

Scale Factor   Transformations and Symmetry

The **scale factor (SF)** is a ratio of the length of one side of a dilated figure to the length of the corresponding side of the original figure. A dilation with a scale factor between 0 and 1, or less than 100%, is a **reduction**. A dilation with a scale factor greater than 1, or 100%, is an **enlargement**.

You can find the scale factor of a dilation by following these steps.

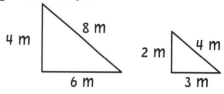

Step 1: Substitute each pair of corresponding sides of the figures into this ratio:

$$\text{scale factor} = \frac{\text{corresponding side length of dilation}}{\text{corresponding side length of original}} \qquad \frac{3}{6} \text{ or } \frac{4}{8} \text{ or } \frac{2}{4}$$

Step 2: Reduce each fraction to its lowest terms. All ratios (fractions) of the corresponding sides should be the same if the image is a dilation.

$$\frac{3}{6} = \frac{1}{2} \qquad \frac{4}{8} = \frac{1}{2} \qquad \frac{2}{4} = \frac{1}{2}$$

$$\text{scale factor} = \frac{1}{2} = 0.5 = 50\%$$

Use the given scale factor to determine whether the dilation is an enlargement or a reduction.

1. SF = $\frac{1}{2}$ _____

2. SF = $\frac{8}{9}$ _____

3. SF = 4 _____

4. SF = 247% _____

5. SF = 120% _____

6. SF = 32% _____

7. SF = $\frac{2}{3}$ _____

8. SF = 8 _____

Find the scale factor of each dilation.

9. 12 in. 12 in.   9 in. 9 in.

12 in. 12 in.   9 in. 9 in.

SF = _____

10. 24 yd. 6 yd.   8 yd. 2 yd.

24 yd. 6 yd.   8 yd. 2 yd.

SF = _____

11.

14 ft. 21 ft.

SF = _____

12.

4 mm   8 mm

SF = _____

Name: _____ Date: _____

Combining Transformations   Transformations and Symmetry

Follow each step.

1. a. Draw a triangle.

b. Rotate the triangle 180° around a point.

c. Reflect the triangle over a horizontal line.

2. a. Draw a square.

b. Dilate the square with a scale factor of $\frac{1}{3}$.

c. Reflect the square over a vertical line.

3. a. Draw a trapezoid.

b. Transform the trapezoid by sliding it to the right 1 inch.

c. Rotate the trapezoid 270° around a point.

4. a. Draw a rhombus.

b. Rotate the rhombus 45° around a point.

c. Transform the rhombus by sliding it vertically $\frac{1}{4}$ of an inch.

Name: _____   Date: _____

Line Symmetry

Transformations and Symmetry

A figure has **line symmetry** if it can be folded along its center line so that the two halves are equal.

These figures have line symmetry.

These figures do not have line symmetry.

Determine if the following figures have line symmetry and write *Yes* or *No*.

1.

_____

2.

_____

3.

_____

4.

_____

5.

_____

6.

_____

7.

_____

8.

_____

9.

_____

10. Draw a figure that is not shown on this page that has at least one line of symmetry. Then, draw the lines of symmetry.

Real-World Line Symmetry   Transformations and Symmetry

Architects use geometric figures when designing buildings, bridges, and homes. List symmetrical figures that architects use. Use the pictures below to help.

_____   _____

_____   _____

_____   _____

_____   _____

_____

Rotational Symmetry   Transformations and Symmetry

A figure has **rotational symmetry** if it can be rotated around its center point between 0° and 360° and appear unchanged. For example, a square has rotational symmetry at 90°, 180°, and 270° clockwise (or 90° counterclockwise).

0°   45°   90°   135°   180°   225°   270°   315°

However, the following triangle does **not** have rotational symmetry. It does **not** appear unchanged at any point between 0° and 360°.

0°   45°   90°   135°   180°   225°   270°   315°

Determine if each figure has rotational symmetry and write *Yes* or *No*.

1.

_____

2.

_____

3.

_____

4.

_____

5.

_____

6.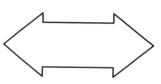

_____

7.

_____

8.

_____

9.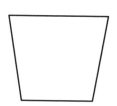

_____

Identifying Tessellations

Regular tessellations occur when congruent regular polygons are repeated, or tiled, to cover a plane without gaps or overlaps. Only squares, hexagons, and triangles may form regular tessellations.

Determine if each set of shapes are regular tessellations and circle *Yes* or *No*.

1.    Yes   No

2.   Yes   No

3. 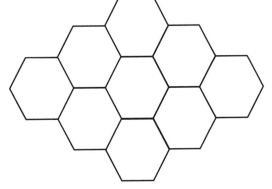   Yes   No

Draw your own regular tessellation using the given regular polygon.

4.

5.

6.

Name: ______________________ Date: ____________

Review

Transformations and Symmetry

Use a word from the box below to complete each statement.

similar	rotational	dilation
line	turn	flip
scale factor	reduces	congruent
enlarges	transformation	tessellations

1. A ______________________ moves a geometric figure from one position to another by flipping, sliding, or turning it.
2. Reflection is another term for a ______________ .
3. The ______________________ is the ratio that describes how a figure's size changes during a dilation.
4. If a figure can be rotated around a point by less than 360° clockwise without it changing, then it has ______________________ symmetry.
5. Rotation is another term for a ______________ .
6. A scale factor of $\frac{1}{2}$ ______________________ the size of a dilated figure.
7. If a figure can be folded in half and have two equal halves, then it has ______________ symmetry.
8. Another word for scaling is ______________ .
9. A scale factor of 120% ______________________ the size of a dilated figure.
10. In a reflection, rotation, or translation, the original figure and the transformed figure are ______________ .
11. When congruent regular polygons are tiled to cover a plane without gaps or overlaps, they form ______________________.
12. In a dilation, the figures are ______________________ but not congruent.

Name: _____   Date: _____

Review   Transformations and Symmetry

Draw the reflection of each figure.

1.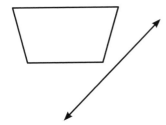

2.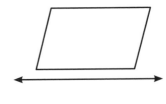

Draw the translation of each figure.

3.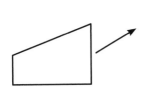

4.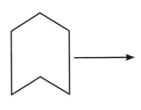

Use a protractor and a ruler to draw each rotation.

5. 90° rotation

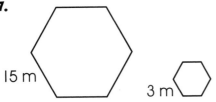

6. 45° rotation

Find the scale factor of each dilation.

7.

15 m   3 m

SF = _____

8.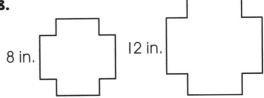

8 in.   12 in.

SF = _____

Venn Diagrams

Venn Diagrams and Coordinate Planes

Venn Diagram—a graphic representation that shows relationships between sets of data

Use words from the box below to complete the Venn diagram.

regular pentagon	isosceles triangle	regular octagon
rectangle	trapezoid	acute triangle
square	right triangle	equilateral triangle

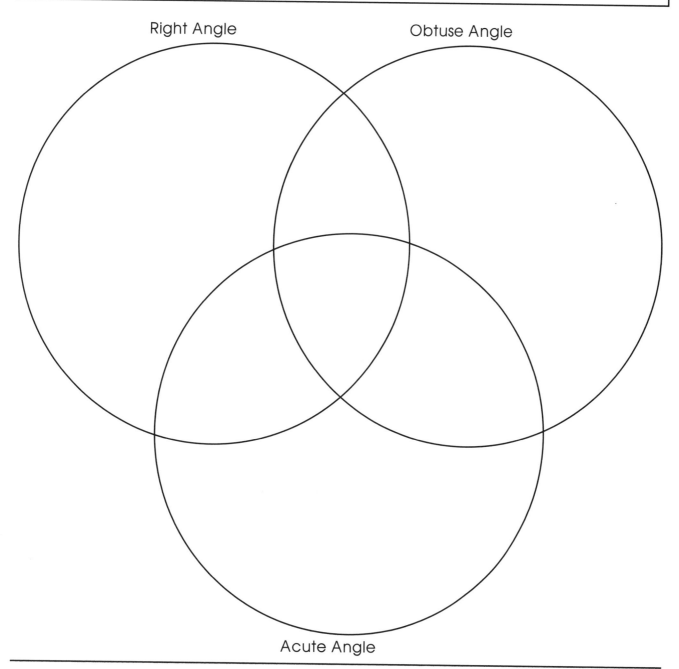

Right Angle   Obtuse Angle

Acute Angle

Venn Diagrams

Venn Diagrams and Coordinate Planes

Use the shapes from the box below to complete the Venn diagram.

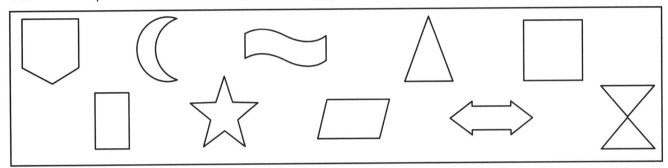

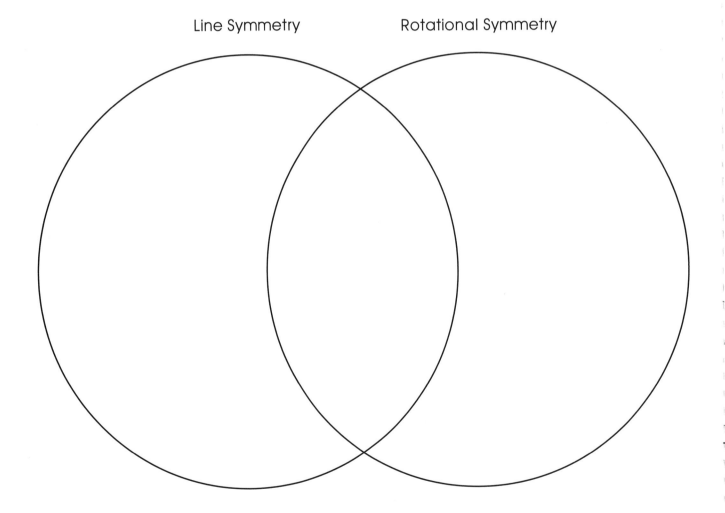

Venn Diagrams

Venn Diagrams and Coordinate Planes

Use the shapes from the box below to complete the Venn diagram.

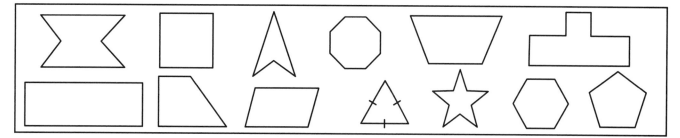

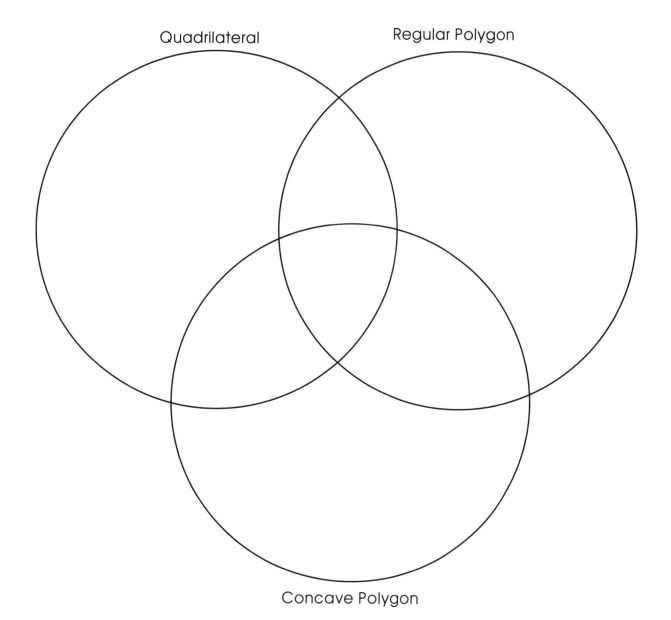

Quadrilateral

Regular Polygon

Concave Polygon

Name: _____   Date: _____

Plotting Points   Venn Diagrams and Coordinate Planes

Coordinate Plane—a plane determined by a horizontal number line (the x-axis) and a vertical number line (the y-axis) that intersect at the **origin**. Each set of coordinates represents an **ordered pair** (x, y). When using a coordinate plane, read the x-coordinate first, then the y-coordinate.

To decide which quadrant an ordered pair is in, determine whether the numbers given are positive or negative. The coordinate plane diagram to the right shows where to plot specific types of ordered pairs.

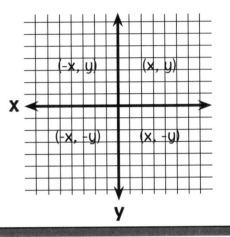

Plot each ordered pair on the corresponding coordinate plane. Connect the points to form a polygon. Then, identify the type of polygon.

1. (-4, 8); (4, 8); (8, 2); (-8, 2); (-4, -4); (4, -4)

2. (-5, 8); (2, 8); (9, 8); (5, -6); (-2, -6); (-9, -6)

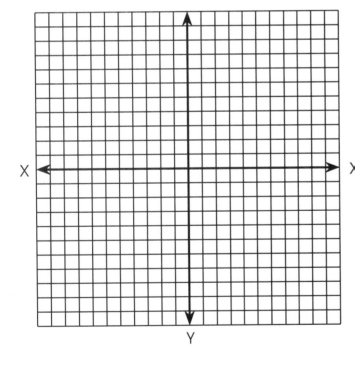

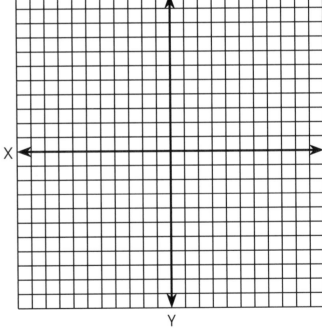

Polygon: _____   Polygon: _____

Coordinate Graphing Basics   Venn Diagrams and Coordinate Planes

Plot, label, and connect the points.

1. A (-7, 6); B (-4, 2); C (-1, -2); D (2, -6); E (7, -2); F (4, 4)

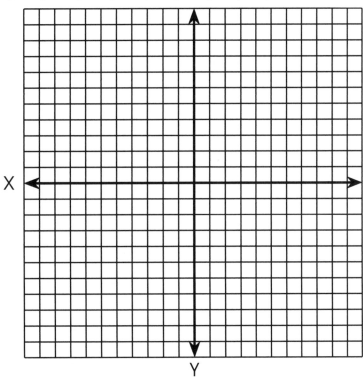

2. A (0, 9); B (3, 9); C (5, 6); D (5, 3); E (3, 0); F (0, 0); G (-2, 3); H (-2, 6)

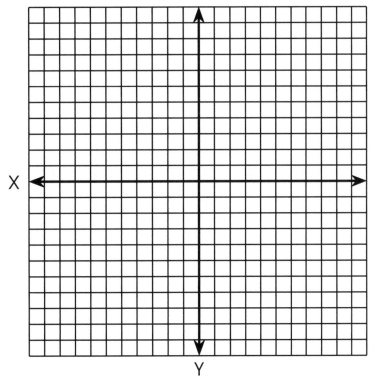

Name: _____ Date: _____

Graphing Polygons — Venn Diagrams and Coordinate Planes

Plot, label, and connect the points in order. Identify each polygon as a rectangle, rhombus, square, or trapezoid.

1. A (2, 5); B (-4, 5); C (-4, -7); D (2, -7)

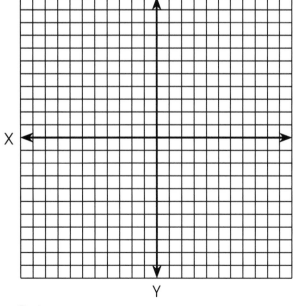

Polygon: _____

2. L (-3, 6); M (0, 9); N (3, 6); O (0, 3)

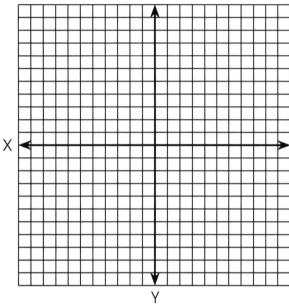

Polygon: _____

3. J (-5, 6); K (-4, -2); L (4, -1); M (3, 7)

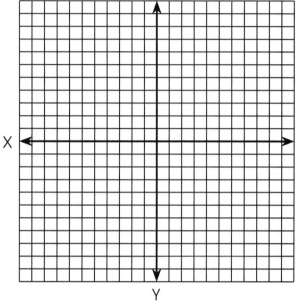

Polygon: _____

4. P (-5, -3); Q (1, -2); R (6, 3); S (7, 9)

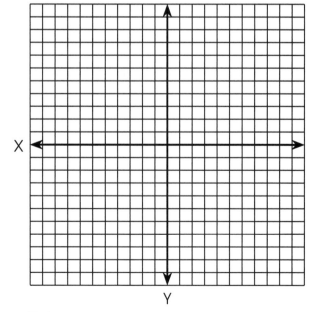

Polygon: _____

Name: _____ Date: _____

Venn Diagrams and Coordinate Planes

Use the key to locate each place in the coordinate plane and record its coordinates.

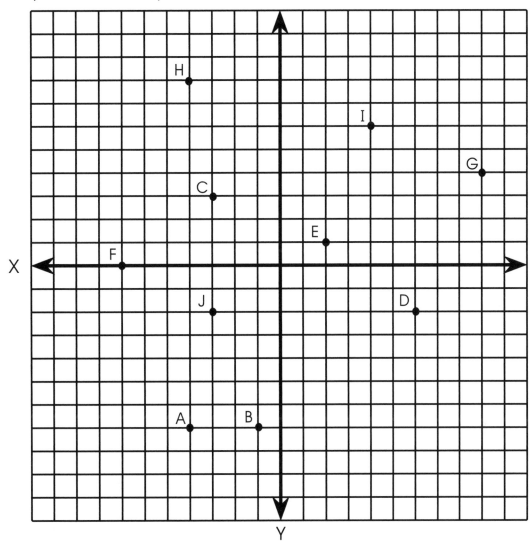

Key

A. Park	D. Grocery Store	G. Theater	J. Restaurant
B. Pool	E. Friend's House	H. Library	
C. School	F. My House	I. Mall	

1. My House   (_____ , _____)   **2.** Park   (_____ , _____)

3. Pool   (_____ , _____)   **4.** Grocery Store   (_____ , _____)

5. School   (_____ , _____)   **6.** Friend's House   (_____ , _____)

7. Mall   (_____ , _____)   **8.** Theater   (_____ , _____)

9. Library   (_____ , _____)   **10.** Restaurant   (_____ , _____)

Name: _____   Date: _____

Riddle

Identify the matching letter for each ordered pair to solve the riddle.

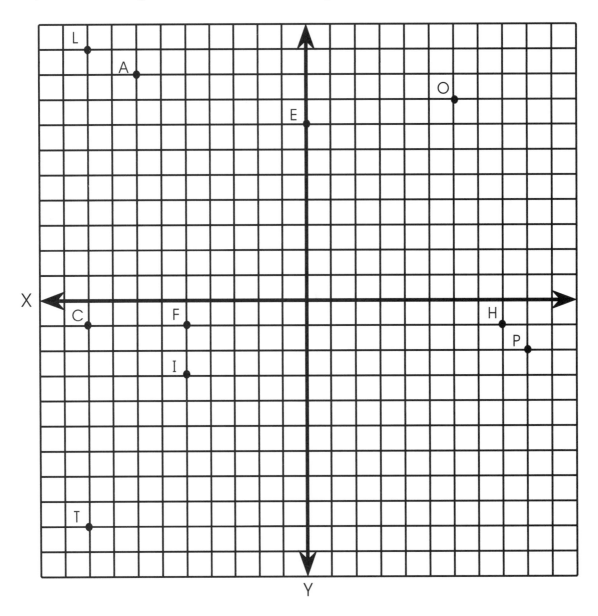

Riddle: What is your math teacher's favorite dessert?

Answer:

$\overline{(-7, 9)}$ $\quad$ $\overline{(9, -2)}$ $\overline{(-5, -3)}$ $\overline{(0, 7)}$ $\overline{(-9, -1)}$ $\overline{(0, 7)}$ $\quad$ $\overline{(6, 8)}$ $\overline{(-5, -1)}$

$\overline{(-9, -1)}$ $\overline{(8, -1)}$ $\overline{(6, 8)}$ $\overline{(-9, -1)}$ $\overline{(6, 8)}$ $\overline{(-9, 10)}$ $\overline{(-7, 9)}$ $\overline{(-9, -9)}$ $\overline{(0, 7)}$ $\quad$ $\overline{(9, -2)}$ $\overline{(-5, -3)}$

Name: _____   Date: _____

Plotting Points Practice   Venn Diagrams and Coordinate Planes

Plot, label, and connect the points in order.

A (0, -10); B (3, -6); C (6, 0); D (7, -1); E (8, 1); F (7, 4); G (5, 7); H (3, 8); I (1, 9); J (-1, 9);
K (-3, 8); L (-5, 7); M (-7, 4); N (-8, 1); O (-7, -1); P (-6, 0); Q (-3, -6)

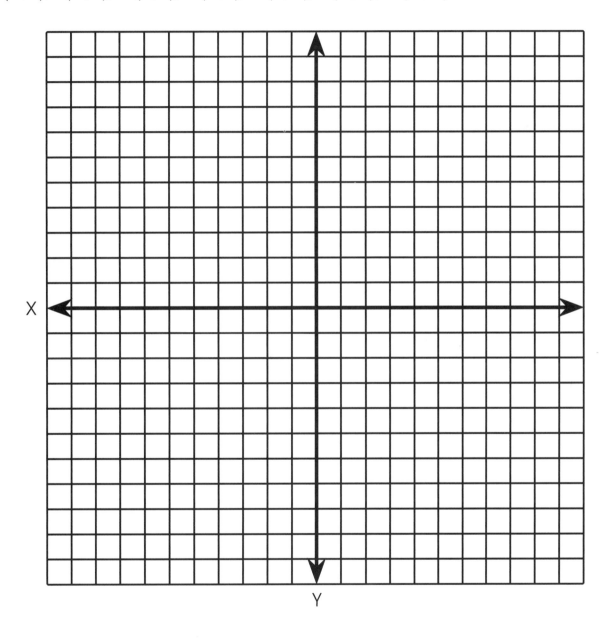

Name: _____ Date: _____

Finding Slope

Venn Diagrams and Coordinate Planes

Slope—the measure of the steepness or incline of a line

You can find the slope of a line by using the coordinates of any two points it passes through.

$\text{slope} = \dfrac{\text{rise}}{\text{run}} = \dfrac{\Delta y}{\Delta x} = \dfrac{y_2 - y_1}{x_2 - x_1}$ (where Δ = change in,

y_2 and y_1 = y-coordinates, and x_2 and x_1 = x-coordinates)

The line passes through the points (1, 5) and (6, 1).

$\text{slope} = \dfrac{\Delta y}{\Delta x} = \dfrac{1 - 5}{6 - 1} = -\dfrac{4}{5}$

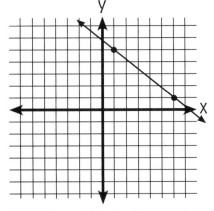

Plot each pair of points and draw a line through them. Then, find each slope. Reduce each slope to its lowest terms.

1. (3, 4); (0, 0)

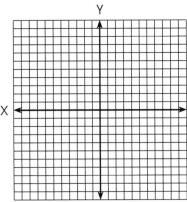

slope = _____

2. (2, 5); (1, 4)

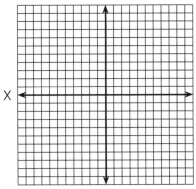

slope = _____

3. (0, 4); (-6, -8)

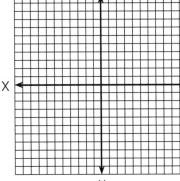

slope = _____

4. (-5, 9); (-1, 1)

slope = _____

5. (-4, 5); (-8, 3)

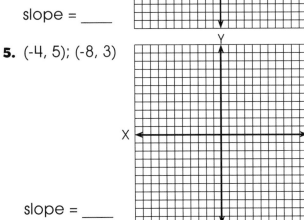

slope = _____

6. (1, 5); (-3, -3)

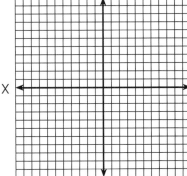

slope = _____

Slope Practice   Venn Diagrams and Coordinate Planes

$$\text{slope} = \frac{\text{rise}}{\text{run}} = \frac{\Delta y}{\Delta x} = \frac{y_2 - y_1}{x_2 - x_1}$$

Use the coordinates provided to find each slope. Show your work.

1. (2, 0); (7, 2)

slope = _____

2. (-4, -2); (-7, -3)

slope = _____

3. (1, 5); (2, -2)

slope = _____

4. (-5, 4); (-3, 3)

slope = _____

5. (8, 6); (6, 8)

slope = _____

6. (5, -5); (2, -1)

slope = _____

7. (7, 9); (-5, -6)

slope = _____

8. (-8, -10); (-4, -4)

slope = _____

9. (12, 10); (1, 4)

slope = _____

10. (-7, 3); (-5, -6)

slope = _____

Name: _____   Date: _____

Mapping Hidden Treasure   Venn Diagrams and Coordinate Planes

Use coordinate graphing to help the explorers create a map to the treasure.

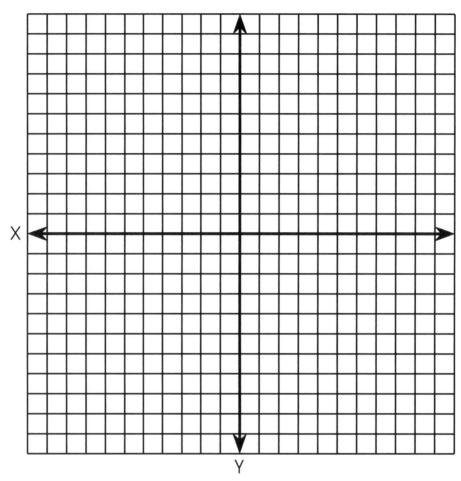

1. Plot and connect the following coordinates in order to show the outline of the island.

   (0, -11); (6, -8); (11, -7); (10, -5); (10, -3); (10, 2); (11, 4); (9, 10); (6, 8); (4, 10); (2, 10);
   (0, 8); (-5, 10); (-10, 5); (-8, 3); (-7, 2); (-8, 0); (-9, -1); (-10, -2); (-6, -5); (-8, -8); (-3, -6);
   (0, -7); (0, -11)

2. Plot and connect the following coordinates to show locations on the island.

   mountain range: (-7, 5); (-6, 3); (-3, 3); (-2, 1); (0, 2); (1, 0); (3, 1); (4, -1); (5, 0);
   river: (5, 1); (7, -2); (5, -3); (5, -5); (4, -7); quicksand: (8, 5); waterfall: (-1, 5);
   lagoon: (-6, -6); cave: (1, 3); buried treasure, point x: (1, -3)

3. Plot and connect the coordinates of this trail from the lagoon to the buried treasure.

   (-6, -6); (-5, -2); (-5, 0); (-4, 2); (-3, 1); (2, 5); (1, -3)

Name: _____   Date: _____

Plot, label, and connect the points in order. Then, name each polygon.

1. A (4, -6); B (-3, -2); C (0, 0); D (2, -3)   **2.** L (0, -1); M (5, -1); N (5, 6); O (0, 6)

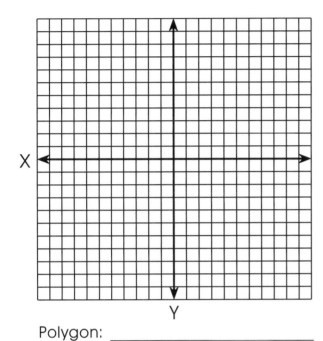

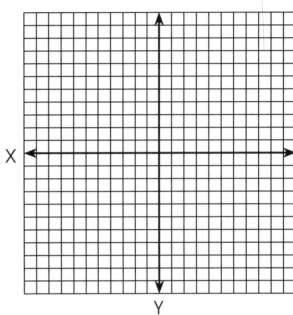

Polygon: _____   Polygon: _____

Plot each pair of points and draw a line through them. Then, find each slope.

3. (-4, -5); (-6, 3)   **4.** (2, 6); (-2, -4)

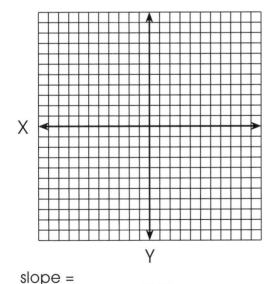

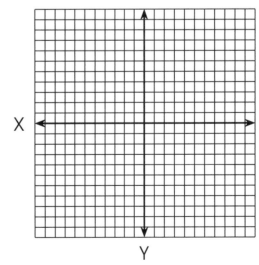

slope = _____   slope = _____

Review   Venn Diagrams and Coordinate Planes

Use the key to locate each place in the coordinate plane. Then, match each place with its ordered pair.

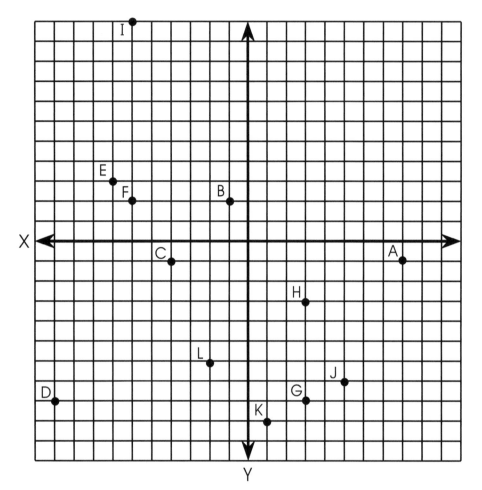

Key

A. Department Store

B. Music Store

C. Sporting Goods Store

D. Jeweler

E. Pet Store

F. Art Gallery

G. Craft and Hobby
   Store

H. Electronics Store

I. Toy Store

J. Restaurant

K. Salon

L. Theater

1. (-7, 3) _____

2. (5, -7) _____

3. (3, -3) _____

4. (8, -1) _____

5. (-2, -6) _____

6. (-4, -1) _____

7. (-6, 2) _____

8. (-6, 11) _____

9. (1, -9) _____

10. (3, -8) _____

11. (-10, -8) _____

12. (-1, 2) _____

Name: _____   Date: _____

Circle the correct answer to each question.

1. a triangle with two 45° interior angles

   a. right   b. acute   c. equiangular   d. obtuse   e. scalene

2. a triangle with a 102° angle

   a. right   b. acute   c. equiangular   d. obtuse   e. scalene

3. a triangle with all angles measuring 60°

   a. right   b. acute   c. equiangular   d. obtuse   e. scalene

Use the diagram to the right to answer each question.

4. If ∠H = 72°,

   ∠I = _____°

   ∠K = _____°

   ∠L = _____°

   ∠M = _____°

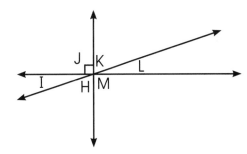

Use the diagram to the right to answer each question.

5. List each pair of corresponding angles.

   _____

6. List each pair of alternate exterior angles.

   _____

7. List each pair of alternate interior angles.

   _____

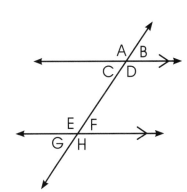

8. List each pair of consecutive interior angles.

   _____

Final Review

Use the diagram to the right to answer each question.

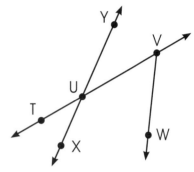

9. Name the point of intersection of the two lines. _____

10. Name three line segments. _____

11. Name three rays. _____

12. Name two sets of collinear points. _____

13. Name the intersecting lines. _____

Determine which rule is used to prove that the triangles below are congruent. Write *ASA*, *SSS*, *AAS*, or *SAS* under each pair of triangles.

14.

15.

16.

_____   _____   _____

Classify each triangle below by its angles and its sides.

17.

18.

19.

_____   _____   _____

_____   _____   _____

On a separate sheet of paper, use the Pythagorean theorem to find the length of each missing side. Round to the nearest hundredth.

20.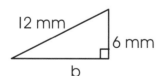
12 mm, 6 mm, b

21.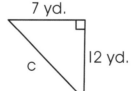
7 yd., 12 yd., c

22.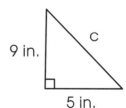
9 in., c, 5 in.

_____   _____   _____

Final Review   Page 3

Use the diagram to the right to tell whether each statement is true or false.

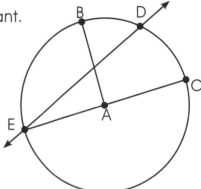

23. _____ Diameter $\overline{EC}$ can also be a secant.

24. _____ $\overleftrightarrow{ED}$ is a secant.

25. _____ $\overset{\frown}{CD}$ is a minor arc.

26. _____ $\overline{DE}$ is a chord.

27. _____ $\overline{AB}$ is a radius.

Find the measure of each arc or angle.

28.

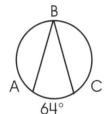

$\angle ABC =$ _____

29.

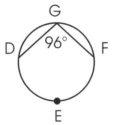

$\overset{\frown}{DEF} =$ _____

30.

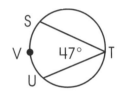

$\overset{\frown}{SVU} =$ _____

Find the area and the perimeter or circumference of each figure. Show your work on a separate sheet of paper.

31.

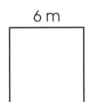
6 m

A = _____ P = _____

32.

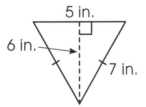

A = _____ P = _____

33.

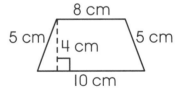

A = _____ P = _____

34.

12 yd.

A = _____ C = _____

35.

8 mm
3 mm

A = _____ P = _____

36.

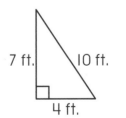

A = _____ P = _____

Use a calculator to find the surface area of each solid. Show your work on a separate sheet of paper. Round to the nearest hundredth.

37.

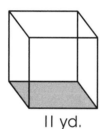

15 mm

8 mm

SA = _____

38.

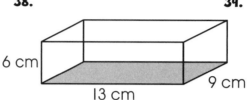

6 cm

13 cm

9 cm

SA = _____

39.

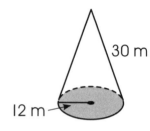

30 m

12 m

SA = _____

40.

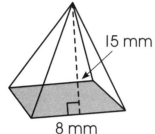

11 yd.

SA = _____

41.

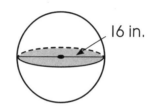

16 in.

SA = _____

42.

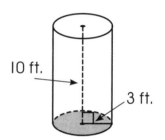

10 ft.

3 ft.

SA = _____

Use a calculator to find the volume of each solid. Show your work on a separate sheet of paper. Round to the nearest hundredth.

43.

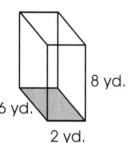

8 yd.

6 yd.

2 yd.

V = _____

44.

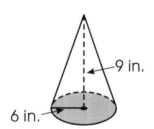

9 in.

6 in.

V = _____

45.

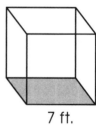

7 ft.

V = _____

46.

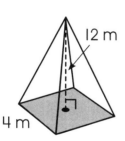

12 m

4 m

V = _____

47.

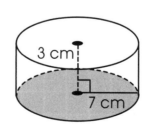

3 cm

7 cm

V = _____

48.

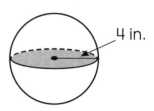

4 in.

V = _____

Final Review

Identify each transformation by writing *reflection*, *rotation*, *translation*, or *dilation*.

49.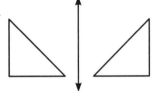

_____

50.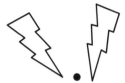

_____

51.

_____

52.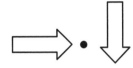

_____

53.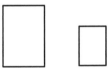

_____

54.

_____

Use a protractor and a ruler to draw each rotation.

55. 270° rotation

56. 45° rotation

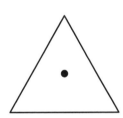

57. 90° rotation

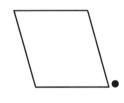

Determine if each figure has rotational symmetry. Write *Yes* or *No*.

58.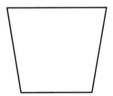

_____

59.

_____

60.

_____

Final Review

Use a word from the box below to identify with each definition.

| slope | x-axis | y-axis | origin | coordinate plane |

61. a plane determined by a horizontal number line and a vertical number line that intersect at the origin _____

62. the measure of the steepness of a line _____

63. horizontal axis _____

64. vertical axis _____

65. where the x-axis and y-axis intersect _____

Plot the following points on the coordinate plane. Then, find the slope for each line below the grid.

A (-7, 2); B (-4, 5); C (-2, 5); D (1, -1); E (7, 7); F (2, 10); G (5, 8); H (2, 6); I (1, -4); J (-2, -5)

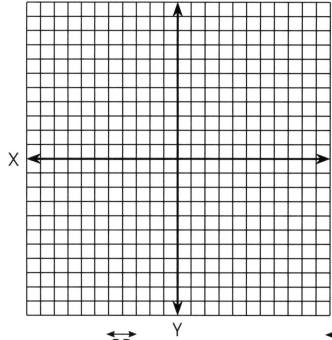

66. $\overleftrightarrow{AB}$ **67.** $\overleftrightarrow{CD}$ **68.** $\overleftrightarrow{EF}$

slope = _____   slope = _____   slope = _____

69. $\overleftrightarrow{GH}$ **70.** $\overleftrightarrow{IJ}$

slope = _____   slope = _____

Geometric Formulas

Pythagorean Theorem

$a^2 + b^2 = c^2$, where a and b are the triangle's legs and c is the hypotenuse

Radius and Diameter of a Circle

Radius: $r = \dfrac{d}{2}$, where r is radius and d is diameter

Diameter: $d = 2r$, where d is diameter and r is the radius

Area

Square: $A = s^2$, where s is the side length

Rectangle: $A = lw$, where l is the length and w is the width

Parallelogram: $A = bh$, where b is the base and h is the height that forms a right angle with the base

Rhombus: $A = \dfrac{1}{2}(d_1 \cdot d_2)$, where d_1 and d_2 are the diagonals

Triangle: $A = \dfrac{1}{2}bh$, where b is the base and h is the height that forms a right angle with the base

Trapezoid: $A = \dfrac{1}{2}h(b_1 + b_2)$, where b_1 and b_2 are the bases and h is the height that forms a right angle with a base

Circle: $A = \pi r^2$, where r is the radius and π is 3.14

Perimeter

Square: $P = 4s$, where s is the side length

Rectangle: $P = 2l + 2w$, where l is the length and w is the width

Parallelogram: $P = 2a + 2b$, where a is one side and b is the other side

Rhombus: $P = 4a$, where a is the side length

Triangle: $P = a + b + c$, where a, b, and c are side lengths

Trapezoid: $P = a + b + c + d$, where a, b, c, and d are side lengths

Circumference

Circle: $C = 2\pi r$, where r is the radius and π is 3.14

$C = \pi d$, where π is 3.14 and d is diameter

Geometric Formulas

Surface Area

Cube: $\qquad$ SA = 6B, where B is the area of the base

Rectangular Prism: $\qquad$ SA = 2B + Ph, where B is the area of the base, P is the perimeter of the base, and h is the height

Square Pyramid: $\qquad$ SA = B + $\frac{1}{2}$Pl, where B is the area of the base, P is the perimeter of the base, and l is the slant height

Cylinder: $\qquad$ SA = 2B + Ch, where B is the area of the base, C is the circumference of the base, and h is the height

Cone: $\qquad$ SA = B + $\frac{1}{2}$Cl, where B is the area of the base, C is the circumference of the base, and l is the slant height

Sphere: $\qquad$ SA = 4πr, where π is 3.14 and r is the radius

Volume

Cube: $\qquad$ V = a^3, where a is the side length

Rectangular Prism: $\qquad$ V = Bh, where B is the area of the base and h is the height

Square Pyramid/Cone: $\qquad$ V = $\frac{1}{3}$Bh, where B is the area of the base and h is the height

Cylinder: $\qquad$ V = Bh, where B is the area of the base and h is the height

Sphere: $\qquad$ V = $\frac{4}{3}\pi r^3$, where r is the radius

Scale Factor

$$SF = \frac{\text{corresponding side length of dilation}}{\text{corresponding side length of original}}$$

Slope

Slope: $\qquad$ Slope = $\dfrac{\text{rise}}{\text{run}} = \dfrac{\Delta y}{\Delta x} = \dfrac{y_2 - y_1}{x_2 - x_1}$, where Δ is the change in, y_2 and y_1 are y-coordinates, and x_2 and x_1 are x-coordinates

Rules

Flat-Plane Rule
If three points are in the same plane, then the line containing two of the points is in the same plane.

Linear Pair Rule
If two angles form a linear pair, then they are supplementary.

Parallel Lines Cut by a Transversal Rules
1. Alternate exterior angles are congruent.
2. Alternate interior angles are congruent.
3. Corresponding angles are congruent.
4. Consecutive interior angles are supplementary.

Parallelogram Rules
1. Both pairs of opposite sides in a parallelogram are parallel.
2. Both pairs of opposite sides in a parallelogram are congruent.
3. Both pairs of opposite angles in a parallelogram are congruent.
4. The diagonals of a parallelogram bisect, or divide, each other into two congruent parts.
5. All consecutive angles (angles that share a side) are supplementary.

Angle-Side-Angle Rule (ASA)
Two triangles are congruent if two angles and the side between those angles of one triangle are congruent to the corresponding parts of the other triangle.

Side-Side-Side Rule (SSS)
Two triangles are congruent if three sides of one triangle are congruent to the corresponding sides of the other triangle.

Angle-Angle-Side Rule (AAS)
Two triangles are congruent if two angles and a side opposite one of those angles of one triangle are congruent to the corresponding parts of the other triangle.

Side-Angle-Side Rule (SAS)
Two triangles are congruent if two sides and the angle between those sides of one triangle are congruent to the corresponding parts of the other triangle.

Congruent Triangle Rules
1. If two sides of a triangle are congruent, then the angles opposite them are also congruent.
2. If two angles of a triangle are congruent, then the sides opposite them are also congruent.
3. If a triangle is equilateral, then it is also equiangular.
4. If a triangle is equiangular, then it is also equilateral.

Measures of Central Angles Rule
If an arc is intercepted, then its measure is equal to the measure of the central angle across from it.

Measures of Inscribed Angles Rule
If an angle is inscribed in a circle, then its measure is equal to half the measure of its intercepted arc.

Glossary

acute angle—an angle with a measure greater than 0° but less than 90°

acute triangle—a triangle with three acute angles

adjacent angles—two angles that have a side and vertex in common

adjacent sides—two sides of a polygon that share a common vertex

alternate exterior angles—pairs of angles formed when a transversal intersects two parallel lines that lie outside the parallel lines on opposite sides of the transversal

alternate interior angles— pairs of angles formed when two parallel lines are intersected by a transversal that lies between the parallel lines on opposite sides of the transversal

angle—two rays that share an endpoint

arc—a portion of a circle's circumference

area—the measure of space within a two-dimensional figure; measured in square units

base—the bottom side of a geometric figure from which the height is drawn

center—the point in a circle that is equidistant from all points on the circle

central angle—an angle in which the vertex is the center of a circle

chord—a line segment that joins two points on a circle

circle—the set of all points in one plane that are equidistant from a fixed point (the center)

circumference—the perimeter of a circle

collinear points—three or more points that lie on the same line

complementary angles—two angles in which the measures add to 90°

concave polygon—a polygon that contains at least one interior angle with a measure greater than 180°; it has lines that intersect within the figure when extended

cone—a solid with one vertex joined to a circular base

congruent—having exactly the same size and shape

congruent angles—angles whose measures are the same

consecutive angles—angles that share a side

consecutive interior angles—pairs of angles formed when a transversal intersects two parallel lines that lie between the parallel lines on the same side of the transversal

convex polygon—a polygon that contains angles whose measures are less than 180°; it has no lines that intersect within the shape when extended

coordinate plane—a plane determined by a horizontal number line (the x-axis) and a vertical number line (the y-axis) that intersect at the origin

coplanar points—three or more points that lie in the same plane

corresponding angles—pairs of angles formed when a transversal intersects two parallel lines that appear in corresponding positions in the two sets of angles

cube—a solid in which every face is a square

cylinder—a solid whose bases are congruent, parallel circles

Glossary

decagon—a 10-sided polygon

diagonal—a line segment that connects two nonconsecutive vertices of a polygon

diameter—a line segment that joins two points on a circle and passes through the center

dilation—a transformation that changes the size of a figure without changing its shape; also known as a scaling

edge—a line segment where two faces in a solid meet

enlargement—a dilation with a scale factor greater than 1 or 100%

equiangular triangle—a triangle that has three congruent angles; each angle measures 60°

equilateral triangle—a triangle that has three congruent sides

face—a shape bounded by edges in a solid

height—the vertical length of a figure or solid from top to bottom

heptagon—a seven-sided polygon

hexagon—a six-sided polygon

hypotenuse—the side opposite the right angle in a right triangle

inscribed angle—an angle formed by two intersecting chords whose vertex is on the circle

intercepted arc—an arc of a circle that lies across from a central angle or an inscribed angle

intersecting lines—lines that cross each other at exactly one point

isosceles triangle—A triangle that has two congruent sides

legs—the sides adjacent to the right angle in a right triangle or the two congruent sides of an isosceles triangle

line—a set of points in a straight path that extends infinitely in both directions

linear pair—a pair of adjacent angles that forms when two lines intersect

line of symmetry—line that divides a two-dimensional figure into two congruent parts

line segment—a finite portion of a line that has two endpoints

line symmetry—the type of symmetry a figure has if it can be folded along its center line so that the two halves are congruent

major arc—an arc created by a central angle whose measure is greater than 180°

midpoint—a point that bisects, or divides, a line segment into two congruent parts

minor arc—an arc created by a central angle whose measure is less than or equal to 180°

nonagon—a nine-sided polygon

obtuse angle—an angle whose measure is greater than 90° but less than 180°

obtuse triangle—a triangle that has one obtuse angle

octagon—an eight-sided polygon

Glossary

opposite rays—Two rays that share an endpoint and extend in opposite directions to form a line

ordered pair—a set of coordinates (x, y) in a coordinate plane

origin—the point at which the x- and y-axis intersect in a coordinate plane; (0, 0)

parallel lines—lines in the same plane that never intersect

parallelogram—a quadrilateral in which opposite sides are parallel

pentagon—a five-sided polygon

perimeter—the measure of distance around a two-dimensional figure

perpendicular lines—two lines that form a right angle at their point of intersection

pi (π)—the ratio of a circle's circumference to its diameter; approximately 3.14

plane—a flat surface that extends infinitely in all directions

point—a position in a plane or in space that has no dimensions

point of intersection—the point at which two lines cross

polygon—a simple, closed plane figure formed by line segments with two sides meeting at each vertex

prism—a solid figure in which two bases are congruent and parallel to one another and in which sides are rectangles

pyramid—a solid figure that has a polygon base and triangular faces that meet at a common vertex

Pythagorean theorem—the formula for finding the length of a missing side in a right triangle: $a^2 + b^2 = c^2$

quadrilateral—a four-sided polygon

radius—a line segment that runs from the center of a circle to a point on the circle

ray—a portion of a line that extends from one endpoint infinitely in one direction

rectangle—a parallelogram that has four right angles

rectangular prism—a prism that has two rectangular bases and four rectangular faces

reduction—a dilation with a scale factor between 0 and 1, or less than 100%

reflection—a transformation that produces a mirror image of a figure reflected across a fixed line; also known as a flip

regular polygon—a polygon that is equilateral (has equal sides) and equiangular (has equal angles)

regular tessellation—a covering of a plane by repeated or tiled regular polygons without overlaps or gaps; only squares, hexagons, and triangles may form regular tesselations

rhombus—a parallelogram with four congruent sides

right angle—an angle whose measure equals 90°

right triangle—a triangle that has one right angle

rotation—a transformation that turns a figure around a fixed point; also known as a turn

rotational symmetry—the type of symmetry a figure has if it can be rotated around its center point by a degree measure greater than 0° and less than 360° and appear unchanged

scale factor (SF)—a ratio of the length of one side of a dilated figure to the length of the corresponding side of the original figure

scalene triangle—a triangle with no congruent sides

secant—a line that intersects a circle at two points

side—an edge or boundary of a geometric figure

similar—two figures that have the same shape but are not the same size

slant height—the distance from the base on a solid object's face to the vertex; must form a right angle with the edge of the base

slope—the measure of the steepness or incline of a line

solid—a closed, three-dimensional figure that contains edges, faces, and vertices

sphere—a solid in which all points on the surface are equidistant from the center

square—a parallelogram with congruent sides and congruent angles

square pyramid—a solid that has a square base and triangular faces that meet at a common vertex

straight angle—an angle with a measure equal to 180°, or a straight line

supplementary angles—two angles with measures that add to 180°

surface area—the sum of the areas of all of the faces in a three-dimensional figure

tangent—a line that intersects a circle at one point

three-dimensional—figures that have length, width, and depth

transformation—a movement of specific change to a geometric figure

translation—a transformation that shifts or moves a shape in one direction without rotation; also known as a slide

transversal—a line that intersects two parallel lines to form eight angles

trapezoid—a quadrilateral with only one pair of parallel sides

triangle—a three-sided polygon

triangular prism—a solid that has two triangular bases and three rectangular faces

triangular pyramid—a solid with a triangular base and triangular faces that meet at a common vertex

two-dimensional—figures that have length and width but not depth

Venn diagram—a graphic representation that shows relationships between sets of data

vertex—a point at which two line segments, lines, or rays meet to form an angle; a point on a solid where three or more faces intersect

volume—the amount of space within a three-dimensional figure; measured in cubic units

Answer Key

For all lines, segments, and angles, accept reverse letter order for student answers.

Page 5

1. $\overrightarrow{XY}$; 2. $\overline{DE}$; 3. points S and T; 4. $\overleftrightarrow{WX}$; 5. $\overleftrightarrow{CD}$; 6. $\overleftrightarrow{LM}$; 7. $\overline{JK}$; 8. points R, V, and O; 9. $\overrightarrow{HI}$

Page 6

1.–6.: Drawings will vary; 7. points L, M, N, and O; 8. Answers will vary but may include $\overline{LM}$, $\overline{MN}$, $\overline{OM}$, $\overline{LN}$; 9. $\overrightarrow{ML}$, $\overrightarrow{MO}$, $\overrightarrow{MN}$; 10. $\overleftrightarrow{LM}$, $\overleftrightarrow{LN}$, $\overleftrightarrow{MN}$; 11. points R, S, T, U, and V; 12. Answers will vary but may include $\overline{SU}$, $\overline{SV}$, $\overline{RS}$, $\overline{RT}$, $\overline{ST}$; 13. Answers will vary but may include $\overrightarrow{SR}$, $\overrightarrow{ST}$, $\overrightarrow{SU}$, $\overrightarrow{SV}$, $\overrightarrow{TR}$, $\overrightarrow{RT}$, $\overrightarrow{ST}$; 14. $\overleftrightarrow{RS}$, $\overleftrightarrow{RT}$, $\overleftrightarrow{ST}$

Page 7

1. perpendicular; 2. perpendicular; 3. parallel; 4. point S; 5. $\overleftrightarrow{TR}$, $\overleftrightarrow{QU}$; 6.–9. Drawings will vary.

Page 8

1. point Q; 2. point Y; 3. point K; 4.–6. Drawings will vary.; 7. 6; 8. 6; 9. 12; 10. 6; 11. 12; 12. 24; 13. 5; 14. 5; 15. 10

Page 9

1. $\overrightarrow{IH}$, $\overrightarrow{IJ}$; 2. Answers will vary but may include $\overleftrightarrow{HJ}$, $\overleftrightarrow{HI}$, $\overleftrightarrow{IJ}$; 3. $\overline{ED}$, $\overline{EF}$; 4. Answers will vary but may include $\overline{DF}$, $\overline{DE}$, $\overline{EF}$; 5. $\overrightarrow{IH}$, $\overrightarrow{IK}$; 6. $\overrightarrow{XW}$, $\overrightarrow{XZ}$; 7. $\overrightarrow{ST}$, $\overrightarrow{SU}$; $\overrightarrow{SR}$, $\overrightarrow{SV}$; 8. $\overrightarrow{BA}$, $\overrightarrow{BC}$; 9. $\overrightarrow{NP}$, $\overrightarrow{NM}$; $\overrightarrow{NQ}$, $\overrightarrow{NO}$; 10. $\overrightarrow{DC}$, $\overrightarrow{DE}$

Page 10

1. True, All three points are on the same line; 2. False, Point Z is not on the same line; 3. True, Point K is not on the same line; 4. False, Only two points are on the line.; 5. points A, B, C and points D, B, E; 6. points A and C; 7. Answers will vary but may include

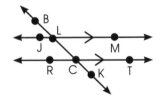

collinear: points J, L, M; points R, C, T; points B, L, C, K; not collinear: Answers will vary.

Page 11

1. No; 2. Yes; 3. Yes; 4. No; 5. not coplanar; 6. coplanar; 7. not coplanar; 8. coplanar

Page 12

1. Yes; 2. No; 3. Yes; 4.–9. Drawings will vary.

Page 13

1. 24 m; 2. $\overleftrightarrow{IH}$, $\overleftrightarrow{IJ}$, $\overleftrightarrow{HJ}$; 3. Answers will vary but may include

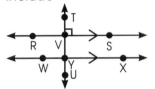

Yes, points W, Y, X, points T, V, Y, U, and points R, V, S; 4. Drawings will vary., Yes

Page 14

1. D; 2. K; 3. I; 4. B; 5. J; 6. H; 7. F; 8. G; 9. C; 10. L; 11. A; 12. E; 13. M; 14. N; 15. B; 16. C; 17. F; 18. A; 19. D; 20. E

Page 15

1. Answers will vary but may include $\overleftrightarrow{KM}$, $\overleftrightarrow{NO}$, $\overleftrightarrow{KL}$, $\overleftrightarrow{NL}$, $\overleftrightarrow{LO}$, $\overleftrightarrow{LM}$; 2. Answers will vary but may include $\overline{KL}$, $\overline{NL}$, $\overline{LO}$, $\overline{LM}$, $\overline{KM}$, $\overline{NO}$; 3. Answers will vary but may include $\overrightarrow{LK}$, $\overrightarrow{LN}$, $\overrightarrow{LO}$, $\overrightarrow{LM}$, $\overrightarrow{KM}$, $\overrightarrow{ON}$, $\overrightarrow{MK}$, $\overrightarrow{NO}$; 4. point L; 5. 9; 6. 18; 7. 9; 8. 9; 9. 18; 10. 36; 11. Answers will vary but may include $\overrightarrow{AB}$ and $\overleftrightarrow{DE}$; 12. points C and F; 13. Answers will vary but may include $\overleftrightarrow{CG}$; 14. points A, C, B; points D, F, E; points C, F, G; 15. yes; 16. no

Page 16

1. straight; 2. acute; 3. right; 4. obtuse; 5. obtuse; 6. acute; 7. acute; 8. obtuse; 9. right; 10. acute; 11. obtuse; 12. straight; 13.–15. Drawings will vary.

Page 17

1. ∠B or ∠ABC; 2. ∠H or ∠GHI; 3. ∠E or ∠DEF; 4. point R, $\overrightarrow{RQ}$ and $\overrightarrow{RS}$; 5. point B, $\overrightarrow{BA}$ and $\overrightarrow{BC}$; 6. point Y, $\overrightarrow{YX}$ and $\overrightarrow{YZ}$; 7. point F, $\overrightarrow{FE}$ and $\overrightarrow{FG}$; 8. point M, $\overrightarrow{ML}$ and $\overrightarrow{MN}$; 9. point U, $\overrightarrow{UT}$ and $\overrightarrow{UV}$; 10. $\overrightarrow{ML}$, $\overrightarrow{MN}$; 11. point M; 12. yes; 13. yes; 14. no, ∠DBE and ∠ABC do not share a side.

Page 18

1. linear pair; 2. FEG; 3. RUV, VUS; RUT, SUT; RUT, RUV; TUS, VUS; 4. 143°; 5. 100°; 6. 22°; 7. 145°, 35°, 145°; 8. 58°, 122°, 58°; 9. 46°, 134°, 46°

Page 19

1. 90°, complementary; 2. A, supplementary; 3. 47°, 137°, 133°; 4. 48°, 42°, 138°; 5. 180°; 6. 180°; 7. 90°; 8. 58°, 122°, 148°; 9. 45°, 135°, 45°; 10. 180°; 11. 90°; 12. complementary angles

Answer Key

Page 20

1. ∠S and ∠Z, ∠T and ∠Y; 2. ∠U and ∠X, ∠V and ∠W; 3. ∠U and ∠W, ∠V and ∠X; 4. ∠S and ∠W, ∠T and ∠X, ∠U and ∠Y, ∠V and ∠Z; 5. corresponding; 6. alternate exterior; 7. alternate interior; 8. consecutive interior; 9. alternate exterior; 10. corresponding

Page 21

1. P and M; 2. O and N; 3. M; 4. M and P; 5. L; 6. a. ∠X and ∠W, Rules 2 and 3; b. ∠S and ∠T, Rules 1 and 3; c. ∠W and ∠X, Rules 1 and 3; d. ∠T and ∠S, Rules 2 and 3; e. ∠V and ∠Y, Rules 2 and 3; f. ∠U and ∠R, Rules 1 and 3; g. ∠U and ∠R, Rules 2 and 3; 7. No, ∠T and ∠X are not consecutive interior angles.; 8. 105°

Page 22

1. acute, 65°; 2. right, 90°; 3. obtuse, 130°; 4. acute, 40°; 5. obtuse, 95°; 6. obtuse, 160°

Page 23

1. H and E; E and H; E (but other answers include A, D, and H); F and G; D (but other answers include A, E, and H); 2. 37°, 53°, 127°; 3. 118°; 4. 115°, 85°; 5. Drawings will vary but include

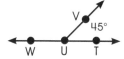

Page 24

1. L; 2. E; 3. I; 4. A; 5. B; 6. F; 7. C; 8. J; 9. H; 10. D; 11. K; 12. G; 13. acute; 14. obtuse; 15. straight; 16. 110°, 70°, 110°; 17. 95°, 85°, 95°; 18. 16°, 164°, 16°

Page 25

1. alternate interior; 2. alternate exterior; 3. consecutive interior; 4. alternate exterior; 5. alternate interior; 6. consecutive interior; 7. acute, 50°; 8. obtuse, 145°; 9. acute, 75°; 10. obtuse, 155°; 11. right, 90°; 12. acute, 60°

Page 26

1. B, D, F; 2. A, all sides are not segments; C, not a closed figure; E, has more than two sides meeting at a vertex; 3. triangle; 4. 4; 5. pentagon; 6. 6; 7. 7; 8. octagon; 9. 9; 10. 10; 11. heptagon; 12. pentagon; 13. nonagon; 14. octagon; 15. hexagon; 16. decagon

Page 27

1. triangle, three sides; 2. quadrilateral, four sides; 3. heptagon, seven sides; 4. pentagon, five sides; 5. nonagon, nine sides; 6. hexagon, six sides; 7. decagon, 10 sides; 8. octagon, eight sides

Page 28

1. equilateral; 2. regular; 3. regular; 4. none; 5. equiangular; 6. regular; 7. regular; 8. none; 9. regular; 10. none; 11. regular; 12. none

Page 29

1. Rule 1; 2. Rule 4; 3. Rules 3 and 4; 4. Rules 3 and 5; 5. Rules 1 and 3; 6. Rule 4; 7. Rule 3; 8. Rules 3 and 5; 9. Rule 1; 10. Rules 1 and 4; 11. Rules 3 and 5; 12. Rules 1 and 3

Page 30

1. No, Only one set of opposite sides is parallel.; 2. Yes; 3. Yes; 4. 7 m, Opposite sides are congruent.; 5. 6 m, Opposite sides are congruent.; 6. 3 m, Diagonals bisect each other.; 7. 65°, Opposite angles are congruent, or consecutive angles are supplementary.; 8. 115°, Opposite angles are congruent, or consecutive angles are supplementary.; 9. 180°, Consecutive angles are supplementary.; 10. Yes, Diagonals bisect each other.

Page 31

1. $\overline{RQ}$; 2. ∠PSR; 3. $\overline{QT}$; 4. $\overline{RT}$; 5. ∠QRS; 6. $\overline{SR}$; 7. 9 ft.; 8. 8 ft.; 9. 7 ft.; 10. 14 ft.; 11. 104°; 12. 76°; 13. 7.5, 15; 14. 30°, 149°; 15. a; 16. c

Page 32

1. concave; 2. convex; 3. concave; 4. concave; 5. convex; 6. convex; 7. convex; 8. concave; 9. concave; 10. convex; 11. concave; 12. convex

Page 33

1. cube; 2. triangular pyramid; 3. triangular prism; 4. sphere; 5. square pyramid; 6.–8. Drawings will vary.

Page 34

1. squares; 2. rectangles; 3. six; 4. one; 5. sphere; 6. 6, 12, 8; 7. 0, 0, 0; 8. 1, 1, 1; 9. 2, 2, 0; 10. 5, 8, 5; 11. 6, 12, 8; 12. cone; 13. cylinder; 14. triangular pyramid; 15. cube; 16. sphere

Page 35

1. triangular prism; 2. rectangular prism; 3. sphere; 4. triangular pyramid; 5. cube; 6. cylinder; 7. cone; 8. square pyramid

Answer Key

Page 36

1. regular; 2. irregular; 3. regular; 4. $\overline{BC}$; 5. ∠ADC; 6. $\overline{BE}$; 7. $\overline{CE}$; 8. ∠BCD; 9. $\overline{DC}$; 10. convex; 11. concave; 12. concave; 13. convex; 14. convex; 15. concave

Page 37

1. 32°, 148°, 148°; 2. $\overline{YX}$; 3. $\overline{VY}$; 4. $\overline{XZ}$; 5. ∠VYX; 6. $\overline{ZW}$ 7. ∠YVW; 8. $\overline{XV}$ and $\overline{WY}$; 9. cube; 10. square pyramid; 11. sphere; 12.–14. Drawings will vary.

Page 38

1. equiangular, equilateral; 2. right, scalene; 3. obtuse, scalene; 4. equiangular, equilateral; 5. acute, scalene; 6. equiangular, equilateral; 7. right, isosceles; 8. acute, isosceles; 9. right, scalene

Page 39

1. D; 2. C; 3. B; 4. E; 5. A; 6. never; 7. always; 8. sometimes; 9. never; 10. always; 11. a; 12. d; 13. c; 14. e; 15. b

Page 40

1. vertex; 2. adjacent sides; 3. hypotenuse; 4. legs; 5. base; 6. base; 7. vertex; 8. adjacent sides; 9. vertex; 10. hypotenuse; 11. vertex; 12. adjacent sides

Page 41

1. ASA; 2. AAS; 3. SSS; 4. SSS; 5. ASA; 6. SAS; 7. AAS; 8. ASA; 9. SAS

Page 42

1. Rule 2; 2. Rule 4; 3. Rule 3; 4. Rule 1; 5. 65°; 6. 19 cm; 7. 19 cm; 8. SUT; 9. 65°; 10. ΔACB

Page 43

1. 78°; 2. 5 mm; 3. 60°; 4. 67°; 5. 70°; 6. 61°; 7. 4 mm; 8. 24 ft.; 9. 54°; 10. 15 in.; 11. 29°; 12. 55°

Page 44

1. 13 m; 2. 12.04 cm; 3. 7.07 yd.; 4. 4.9 ft.; 5. 5.2 mm; 6. 8 in.; 7. 2.24 ft.; 8. 3 cm

Page 45

1. 60°, 60°, 60°, An equilateral triangle is equiangular. The sum of a triangle's measures equals 180°. Therefore, each angle equals 60°; 2. 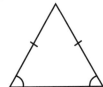 3. 45°, The sum of a triangle's measures equals 180°. 180° – 135° = 45°., 11.31 feet; 4. yes, 65°, 50°, 13 m

Page 46

1. 7.5 cm; 2. 3 yd.; 3. 6.5 ft.; 4. 1 m; 5. 22.5 in.; 6. 10.5 ft.; 7. 2 in.; 8. 5.5 cm; 9. 18 cm; 10. 56 m; 11. 48.5 yd.; 12. 8.5 ft.

Page 47

1. 52 m; 2. 124 ft.; 3. 72 yd.; 4. 128 mm; 5. 26 in.; 6. 66 cm; 7. 38 mm; 8. 40 in.; 9. 14 ft.; 10. 92 yd.; 11. 44 ft.; 12. 134 mm

Page 48

1. $\overline{LM}$; 2. $\overleftrightarrow{TV}$; 3. $\overleftrightarrow{LM}$; 4. No, $\overline{ZS}$ is not a line; 5. No, $\overline{XS}$ is not a line and it intersects the circle at more than one point.; 6. $\overline{CD}$; 7. $\overleftrightarrow{FG}$ and $\overleftrightarrow{HI}$; 8. $\overleftrightarrow{AB}$; 9. No, $\overline{CD}$ is not a line; 10. No, $\overline{ED}$ does not join two points on the circle

Page 49

1. central angle; 2. major arc; 3. minor arc; 4. minor arc; 5. major arc; 6. minor arc; 7. central angle; 8. central angle; 9. central angle; 10. minor arc; 11. major arc; 12. minor arc; 13. major arc; 14. minor arc; 15. major arc

Page 50

1. 64°; 2. 226°; 3. 174°; 4. 180°; 5. 34°; 6. 84°; 7. 76°; 8. 62°; 9. 43°; 10. 58°; 11. 122°; 12. 98°

Page 51

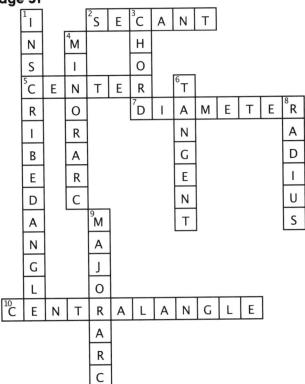

Page 52

1. 17.5 cm; 2. 14.5 yd.; 3. 50 in.; 4. 22 mm; 5. $\overline{AD}$; 6. $\overline{CD}$ or $\overline{AC}$; 7. $\overline{BF}$; 8. $\overleftrightarrow{BF}$; 9. $\overleftrightarrow{GH}$; 10.–15. Drawings will vary.

Page 53

1. chord; 2. tangent; 3. minor arc; 4. secant; 5. major arc; 6. diameter; 7. minor arc; 8. radius; 9. 70°; 10. 240°; 11. 155°; 12. 135°; 13. 37.5°; 14. 120°; 15. 76°; 16. 50°

Page 54

1. 24 cm², 20 cm; 2. 144 in.², 48 in.; 3. 36 ft.², 30 ft.; 4. 16 yd.², 20 yd.; 5. 70 mm², 38 mm; 6. 64 m², 32 m; 7. 16 in.², 16 in.; 8. 72 ft.², 36 ft.; 9. 21 cm², 20 cm

Page 55

1. 180 cm², 58 cm; 2. 100 yd.², 54 yd.; 3. 60 ft.², 36 ft.; 4. 42 m², 30 m; 5. 68 mm², 46 mm; 6. 180 in.², 56 in.; 7. 144 ft.², 52 ft.; 8. 55 cm², 34 cm; 9. 60 m², 36 m

Page 56

1. 120 in.², 52 in.; 2. 30 yd.², 30 yd.; 3. 270 cm², 78 cm; 4. 1,080 ft.², 156 ft.; 5. 367.5 m², 91 m; 6. 750 mm², 130 mm

Page 57

1. 6 yd.², 12 yd.; 2. 35 m², 34 m; 3. 45 in.², 34 in.; 4. 73.5 ft.², 46 ft.; 5. 67.5 cm², 45 cm; 6. 150 yd.², 55 yd.; 7. 21 m², 27 m; 8. 54 cm², 36 cm; 9. 58.5 ft.², 39 ft.

Page 58

1. 40 m², 30 m; 2. 72 in.², 34.5 in.; 3. 35 cm², 40 cm; 4. 77.5 yd.², 44.5 yd.; 5. 64 mm², 42 mm; 6. 192 m², 68 m; 7. 160 ft.²

Page 59

1. 153.86 yd.², 43.96 yd.; 2. 1,962.5 cm², 157 cm; 3. 379.94 ft.², 69.08 ft.; 4. 200.96 yd.², 50.24 yd.; 5. 254.34 mm², 56.52 mm; 6. 314 cm², 62.8 cm

Page 60

1. 64 m², 32 m; 2. 70 mm², 38 mm; 3. 24 yd.², 24 yd.; 4. 120 m², 52 m; 5. 100 yd.², 64 yd.; 6. 52 m², 36 m; 7. 84 cm², 42 cm; 8. 354 yd.², 76 yd.

Page 61

1. 200 m²; 2. 1,240 ft.²; 3. 12 ft.; 4. 270 in.²; 5. w=9 in., P=28 in.; 6. s=25 cm, A=625 cm²; 7. 51 mm²; 8. 75.36 m; 9. 4 cm; 10. 14 in.

Page 62

1. 1,536 mm²; 2. 96 ft.²; 3. 864 cm²; 4. 150 yd.²; 5. 486 in.²; 6. 600 m²; 7. 6 cm²; 8. 1,014 ft.²; 9. 1,350 yd.²

Page 63

1. 496 ft.²; 2. 346 mm²; 3. 880 in.²; 4. 188 cm²; 5. 592 m²; 6. 424 yd.²

Page 64

1. 80 ft.²; 2. 165 cm²; 3. 460 yd.²; 4. 297 m²; 5. 51 in.²; 6. 256 mm²

Page 65

1. 439.6 mm²; 2. 2,260.8 ft.²; 3. 226.08 cm²; 4. 628 ft.²; 5. 715.92 yd.²; 6. 847.8 m²

Page 66

1. 150.72 in.²; 2. 942 m²; 3. 452.16 ft.²; 4. 876.06 ft.²; 5. 489.84 cm²; 6. 967.12 yd.²

Page 67

1. 803.84 ft.²; 2. 314 yd.²; 3. 1,256 in.²; 4. 50.24 cm²; 5. 1,017.36 ft.²; 6. 2,122.64 m²; 7. 2,826 mm²; 8. 1,519.76 in.²; 9. 1,808.64 cm²; 10. 50.24 m²; 11. 2,461.76 ft.²; 12. 452.16 yd.²

Page 68

1. 166 ft.²; 2. 136 cm²; 3. 98 in.²; 4. 256 in.²; 5. 204 cm²; 6. 576 mm²; 7. 602.88 in.²; 8. 301.44 ft.²; 9. 942 yd.²; 10. 103.62 ft.²; 11. 301.44 cm²; 12. 266.9 yd.²

Page 69

1. 125 cm³; 2. 8 ft.³; 3. 343 yd.³; 4. 1,000 mm³; 5. 1,728 in.³; 6. 1 m³; 7. 3,375 ft.³; 8. 216 cm³; 9. 729 mm³

Page 70

1. 240 ft.³; 2. 240 in.³; 3. 40 cm³; 4. 90 cm³; 5. 140 m³; 6. 585 yd.³; 7. 256 m³; 8. 1,404 mm³; 9. 2,520 in.³

Page 71

1. 96 mm³; 2. 53.33 cm³; 3. 277.33 in.³; 4. 65.94 ft.³; 5. 100.48 yd.³; 6. 615.44 cm³

Page 72

1. 1,130.4 ft.³; 2. 254.34 in.³; 3. 1,205.76 cm³; 4. 235.5 ft.³; 5. 138.16 m³; 6. 197.82 m³; 7. 5,837.26 mm³; 8. 1,020.5 cm³; 9. 62.8 yd.³

Answer Key

Page 73

1. 33.49 cm³; 2. 267.95 ft.³; 3. 113.04 yd.³;
4. 65.42 cm³; 5. 2,143.57 m³; 6. 1,436.03 in.³;
7. 4,186.67 mm³; 8. 381.51 ft.³; 9. 904.32 m³

Page 74

1. 50 ft.³; 2. 3,454 ft.²; 3. 72 ft.³; 4. 24,416.64 in.³;
5. 267.95 cm³; 6. 100 m³; 7. 384 mm²; 8. 64 yd.³;
9. 314 cm²; 10. 340 m²

Page 75

1. 6 ft.², 12 ft.; 2. 60 yd.², 34 yd.; 3. 28 m², 24 m;
4. 200.96 ft.², 50.24 ft.; 5. 254.34 mm², 56.52 mm;
6. 30 m², 30 m; 7. 376.8 mm²; 8. 486 in.²; 9. 198 yd.²;
10. 94.2 cm²; 11. 180 mm²; 12. 314 m²

Page 76

1. 150.72 cm³; 2. 512 ft.³; 3. 2,307.9 m³; 4. 300 yd.³;
5. 904.32 in.³; 6. 83.33 mm³; 7. 1,004.8 ft.³;
8. 280 mm³; 9. 80 cm³; 10. 216 m³; 11. 75.36 yd.³;
12. 267.95 m³

Page 77

1. VUT, UVT, UTV; 2. ABC and GHI, BCD and JIH,
CDE and IJF, DEA and JFG, EAB and FGH;
3. LMN and QPR, MNL and PRQ, NLM and RQP;
4. WXY and ABC, XYZ and BCD, YZW and CDA,
ZWX and DAB; 5. EFG, FGH, GHE, HEF; 6. JKL and
OPQ, KLM and PQR, LMN and QRS, MNJ and
RSO, NJK and SOP

Page 78

1. rotation; 2. reflection; 3. none; 4. translation;
5. rotation; 6. translation; 7. none; 8. reflection

Page 79

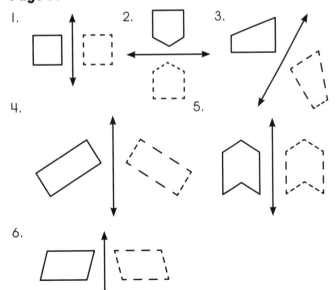

Page 80

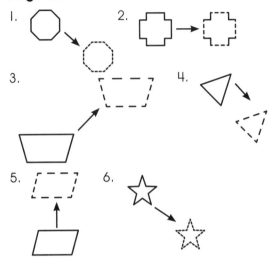

Page 81

1. 270°; 2. 135°; 3. 45°; 4. 0°; 5. 335°; 6. 228°;
7. 310°; 8. 321°

Page 82

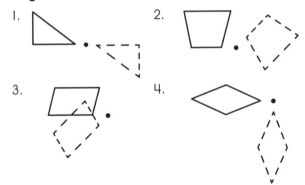

Answer Key

Page 83

1.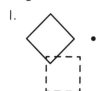

2.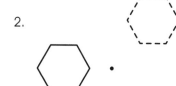

3.

4.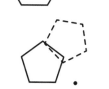

5.

6.

7.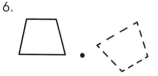

8.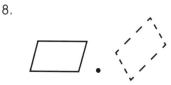

Page 84

1. Yes, They are the same shape but different sizes.; 2. Yes, They are the same shape but different sizes.; 3. Yes, They are the same shape but different sizes.; 4. Yes, They are the same shape but different sizes.; 5. No, The second triangle is rotated, and both triangles are the same size.; 6. Yes, They are the same shape but different sizes.; 7. Yes, They are the same shape but different sizes.; 8. No, The second shape is reflected, and both shapes are the same size.

Page 85

1. reduction; 2. reduction; 3. enlargement; 4. enlargement; 5. enlargement; 6. reduction; 7. reduction; 8. enlargement; 9. $\frac{3}{4}$, 0.75, 75%; 10. $\frac{1}{3}$, 0.3$\overline{3}$, 33%; 11. $\frac{3}{2}$, 1.5, 150%; 12. $\frac{2}{1}$, 2, 200%

Page 86

1.–4. Drawings will vary.

Page 87

1. Yes; 2. Yes; 3. Yes; 4. No; 5. Yes; 6. Yes; 7. No; 8. No; 9. Yes; 10. Drawings will vary.

Page 88

Answers will vary, but may include—triangles, squares, rectangles, prisms, cubes, cylinders, pyramids, and cones.

Page 89

1. No; 2. Yes; 3. No; 4. Yes; 5. Yes; 6. Yes; 7. Yes; 8. No; 9. No

Page 90

1. Yes; 2. No; 3. Yes; 4.–6. Drawings will vary.

Page 91

1. transformation; 2. flip; 3. scale factor; 4. rotational; 5. turn; 6. reduces; 7. line; 8. dilation; 9. enlarges; 10. congruent; 11. tessellations; 12. similar

Page 92

1.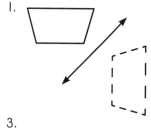

2.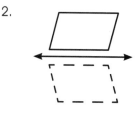

3.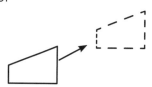

4.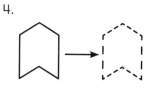

5.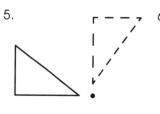

6.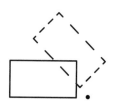

7. $\frac{1}{5}$, 0.20, 20%; 8. $\frac{3}{2}$, 1.5, 150%

Answer Key

Page 93

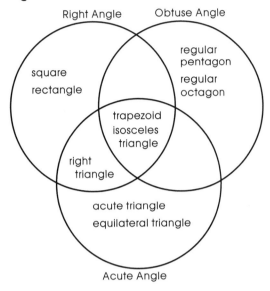

Page 94

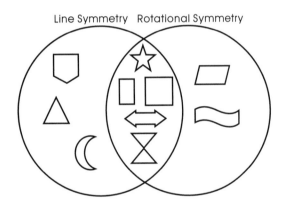

Page 95

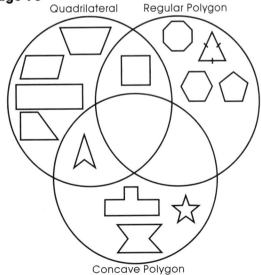

Page 96

1.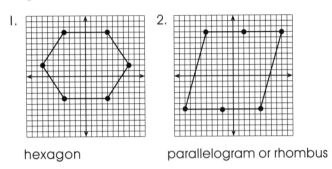

hexagon

2. parallelogram or rhombus

Page 97

1.
2.

Page 98

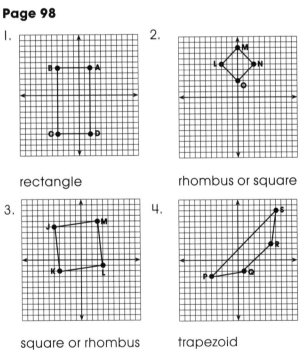

1. rectangle
2. rhombus or square
3. square or rhombus
4. trapezoid

Page 99

1. (-7, 0); 2. (-4, -7); 3. (-1, -7); 4. (6, -2); 5. (-3, 3);
6. (2, 1); 7. (4, 6); 8. (9, 4); 9. (-4, 8); 10. (-3, -2)

Page 100

A PIECE OF CHOCOLATE PI

Answer Key

Page 101

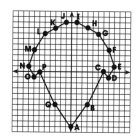

Page 102

1.
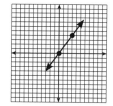
slope = $\frac{4}{3}$

2.
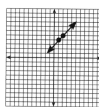
slope = 1

3.
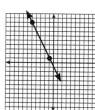
slope = 2

4.
slope = –2

5.
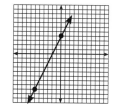
slope = $\frac{1}{2}$

6.
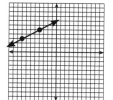
slope = 2

Page 103

1. $\frac{2}{5}$; 2. $\frac{1}{3}$; 3. –7; 4. $-\frac{1}{2}$; 5. –1; 6. $-\frac{4}{3}$; 7. $\frac{5}{4}$; 8. $\frac{3}{2}$; 9. $\frac{6}{11}$; 10. $-\frac{9}{2}$

Page 104

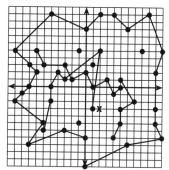

Page 105

1.
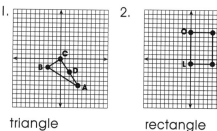
triangle

2.
rectangle

3.
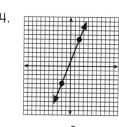
slope = –4

4.
slope = $\frac{5}{2}$

Page 106

1. E. Pet Store; 2. J. Restaurant; 3. H. Electronics Store; 4. A. Department Store; 5. L. Theater; 6. C. Sporting Goods Store; 7. F. Art Gallery; 8. I. Toy Store; 9. K. Salon; 10. G. Craft and Hobby Store; 11. D. Jeweler; 12. B. Music Store

Page 107

1. a; 2. d; 3. c; 4. 18, 72, 18, 90; 5. ∠A and ∠E, ∠B and ∠F, ∠C and ∠G, ∠D and ∠H; 6. ∠A and ∠H, ∠B and ∠G; 7. ∠C and ∠F, ∠D and ∠E; 8. ∠C and ∠E, ∠D and ∠F

Answer Key

Page 108

9. point U; 10. Answers will vary but may include
$\overline{TU}$, $\overline{UV}$, $\overline{UY}$, $\overline{UX}$, $\overline{YX}$, $\overline{TV}$, $\overline{VW}$; 11. Answers will vary but
may include $\overrightarrow{UT}$, $\overrightarrow{UY}$, $\overrightarrow{UX}$, $\overrightarrow{UV}$, $\overrightarrow{VW}$, $\overrightarrow{VT}$, $\overrightarrow{TV}$;
12. points T, U, and V, and points X, U, and Y; 13. $\overleftrightarrow{TV}$
and $\overleftrightarrow{YX}$; 14. AAS; 15. SSS; 16. ASA; 17. equiangular,
equilateral; 18. acute, isosceles; 19. right, scalene;
20. 10.39 mm; 21. 13.89 yd.; 22. 10.30 in.

Page 109

23. false; 24. true; 25. true; 26. true; 27. true;
28. 32°; 29. 192°; 30. 94°; 31. 36 m², 24 m;
32. 15 in.², 19 in.; 33. 36 cm², 28 cm; 34. 113.04 yd.²,
37.68 yd.; 35. 24 mm², 22 mm; 36. 14 ft.², 21 ft.

Page 110

37. 304 mm²; 38. 498 cm²; 39. 1,582.56 m²;
40. 726 yd.²; 41. 803.84 in.²; 42. 244.92 ft.²;
43. 96 yd.³; 44. 339.12 in.³; 45. 343 ft.³; 46. 64 m³;
47. 461.58 cm³; 48. 267.95 in.³

Page 111

49. rotation; 50. rotation; 51. translation;
52. reflection; 53. dilation; 54. dilation;

55.    56.

57. 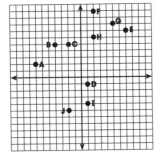   58. No; 59. No; 60. Yes

Page 112

61. coordinate plane; 62. slope; 63. x-axis;
64. y-axis; 65. origin;

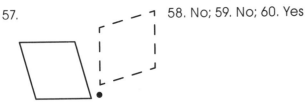

66. 1; 67. –2;

68. $-\frac{3}{5}$; 69. $\frac{2}{3}$;

70. $\frac{1}{3}$